Cenozoic Fossils II
The Neogene

Bruce L. Stinchcomb

4880 Lower Valley Road, Atglen, Pennsylvania 19310

Dedication

This work deals with that portion of geologic time and history closest to the present—the Neogene consisting of the Miocene, Pliocene, Pleistocene, and Holocene Epochs. As a major facilitator in this work, I would like to dedicate it to my beloved wife Karoline Stinchcomb (1935-2009). Karoline, besides assisting in multiple ways with my paleontological interests, was instrumental in arranging trips to places that gave an opportunity to geologize in Neogene strata—specifically the Miocene molasse near Ulm and the Steinheim Astroblem, both in southern Germany. Also arranged and facilitated by Karoline were delightful trips to the Caribbean where fossils and strata of the Pliocene and Pleistocene epochs were extensively experienced. Some of the results of these experiences geologizing in late Cenozoic rocks are incorporated in this work.

Other Schiffer Books By The Author:
Cenozoic Fossils I: Paleogene. ISBN: 9780764334245. $29.99
Mesozoic Fossils: Triassic and Jurassic. ISBN: 9780764331633. $29.99
Mesozoic Fossils II: The Cretaceous Period. ISBN: 9780764332593. $29.99
Paleozoic Fossils. ISBN: 9780764329173. $29.95
World's Oldest Fossils. ISBN: 9780764326974. $29.95

Copyright © 2010 by Bruce L. Stinchcomb

Library of Congress Control Number: 2009939149

All rights reserved. No part of this work may be reproduced or used in any form or by any means—graphic, electronic, or mechanical, including photocopying or information storage and retrieval systems—without written permission from the publisher.
The scanning, uploading and distribution of this book or any part thereof via the Internet or via any other means without the permission of the publisher is illegal and punishable by law. Please purchase only authorized editions and do not participate in or encourage the electronic piracy of copyrighted materials.
"Schiffer," "Schiffer Publishing Ltd. & Design," and the "Design of pen and inkwell" are registered trademarks of Schiffer Publishing Ltd.

Designed by Mark David Bowyer
Type set in Benguiat Bk BT / Aldine721 BT

ISBN: 978-0-7643-3580-8
Printed in China

Schiffer Books are available at special discounts for bulk purchases for sales promotions or premiums. Special editions, including personalized covers, corporate imprints, and excerpts can be created in large quantities for special needs. For more information contact the publisher:

Published by Schiffer Publishing Ltd.
4880 Lower Valley Road
Atglen, PA 19310
Phone: (610) 593-1777; Fax: (610) 593-2002
E-mail: Info@schifferbooks.com

For the largest selection of fine reference books on this and related subjects, please visit our web site at
www.schifferbooks.com
We are always looking for people to write books on new and related subjects. If you have an idea for a book please contact us at the above address.

This book may be purchased from the publisher.
Include $5.00 for shipping.
Please try your bookstore first.
You may write for a free catalog.

In Europe, Schiffer books are distributed by
Bushwood Books
6 Marksbury Ave.
Kew Gardens
Surrey TW9 4JF England
Phone: 44 (0) 20 8392 8585; Fax: 44 (0) 20 8392 9876
E-mail: info@bushwoodbooks.co.uk
Website: www.bushwoodbooks.co.uk

Contents

Acknowledgments & Introduction _____ 4

Chapter One. Cenozoic-II The Neogene _____ 5

Tertiary Period

Chapter Two. Plants _____ 25

Chapter Three. Miocene and Pliocene Corals and Mollusks _____ 41

Chapter Four. Arthropods and Echinoderms _____ 52

Chapter Five. Sharks, Rays, and Fish _____ 61

Chapter Six. Neogene Amphibians and Reptiles _____ 73

Chapter Seven. Miocene and Pliocene Mammals _____ 80

Quaternary Period

Chapter Eight. Invertebrates of the Pleistocene Epoch _____ 86

Chapter Nine. Pleistocene Insects _____ 104

Chapter Ten. Pleistocene Vertebrates _____ 120

Chapter Eleven. The Holocene or Recent Epoch _____ 150

Glossary _____ 158

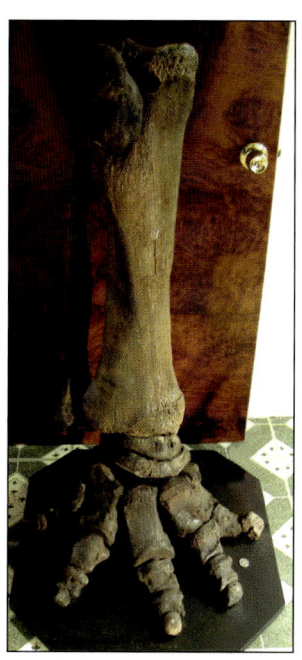

Acknowledgments

The author acknowledges the following persons for a variety of forms of assistance in the culmination of this work. For photos of ice age specimens, thanks to Richard Hagar and John Stade—both intensive workers with Pleistocene fossils. Acknowledgements also to Nancy and the late Roland Kirsten of Ulm, Germany, the late Peter Beirweiler of Echlishausen, R. Kuhn of Gunzburg, Germany, Gerd Muller of Steinheim, Germany, Lisa and Eugen Hascher of Giengen Brenz, Germany, and Piero Garonetti, Pavia, Italy. Thanks also to Brent Ashcroft, Jim Houser, Steve Riggs Jones, Don McKinnis, and Bill Teeters for various assists. For cover art, thanks to Elizabeth V. Stinchcomb. The author also wishes to thank and acknowledge assists in various ways from a wide range of associates who over the years have contributed in many ways toward the culmination of this and other works of this series.

Introduction

This book covers the latest parts of geologic time: the Miocene, Pliocene, Pleistocene, and Holocene Epochs. This is that time represented by fossils of the Neogene, the youngest half of the Cenozoic Era. The Neogene, in many ways, was a peculiar part of geologic time. In its earliest part, there existed the warm conditions of the first half of the Cenozoic Era—the Paleogene. The latter part of the Neogene finds a cooling earth culminating in the ice age (or ice ages), that portion of geologic time representing a unique segment of earth's history. Most significantly, however, it was in the Neogene that the phenomenon of human intelligence appeared—an epochal phenomenon that actually made part of the Universe conscious of itself.

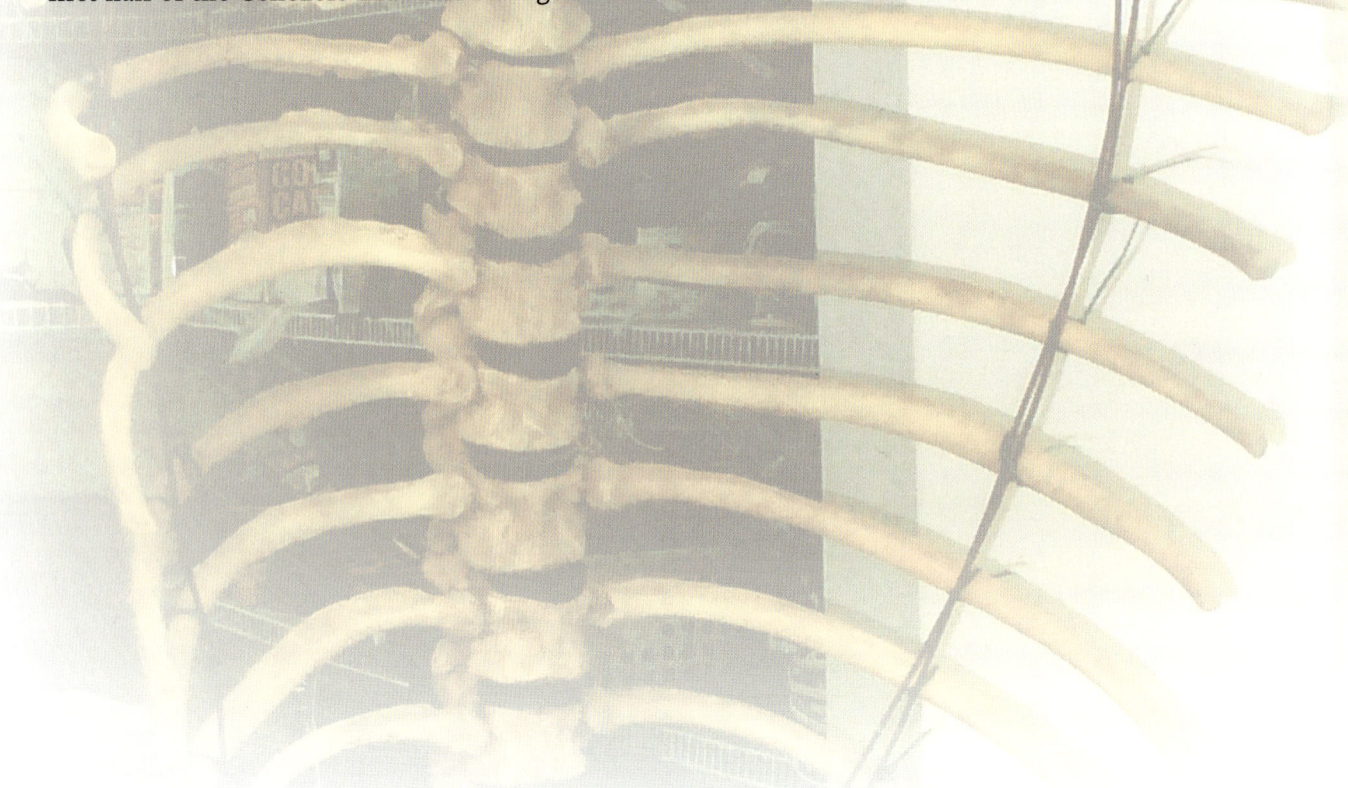

Chapter One
Cenozoic II
The Neogene

Introduction

The Neogene, the last and youngest half of the Cenozoic Era, commences some twenty-eight million years ago with the Miocene Epoch. This geological benchmark is **the real beginning of the modern world!** Early Cenozoic fossils (and the life forms they represent) were similar in some ways to today's life—but it is a similarity "flavored" in an archaic way. Present day animals and plants, those familiar to us today, become the "order-of-the-day" in the Miocene Epoch. This includes such groups as mammals and various angiosperms (especially the grasses), as well as marine mollusks—the fossil shells of which originally were used to delineate the epochs of the Cenozoic Era by Charles Lyell. Lyell's original determination of what constituted the Miocene Epoch proposed that the majority (seventy percent) of its molluscan genera and seventeen percent of its species be living today. The Pliocene Epoch, the second epoch of the Neogene, introduces an even more modern flora and fauna—again originally determined from the percentage of living species found in its rocks. According to Lyell's original concept, Cenozoic marine rocks, which have around fifty percent modern molluscan species, belong to the Pliocene. When Lyell proposed this scheme he was (somewhat) unaware of the extent of the ice age, a more complete understanding of which would come later, predominantly in the late nineteenth and early twentieth centuries.

The Neogene, the youngest half of the Cenozoic Era, records some puzzling aspects of the world's climate—the climate of the last thirty million years. World climate in the Miocene was similar to that prior to that epoch—that is warm and tropical at lower latitudes and temperate at higher ones. **Cold environments appear to have been a rarity**. The Pliocene appears to have been somewhat cooler, if the fossil record is being read correctly; but the Pliocene still favored a warm and equitable climate for the most part. During the Miocene and Pliocene epochs, tropical animals and plants flourished at higher latitudes than they do today. For instance, the large tortoises characteristic of these two epochs, could not live in Texas or in Arkansas today as the winters there are currently too cold. This is because large tortoises, like those found as fossils, cannot dig deep enough into the ground to hibernate when it gets below freezing, as can the smaller box tortoises found there today.

The latter part of the Pliocene saw a cooling of local temperatures and with the beginning of the Pleistocene Epoch (approx. 2.5 million years ago) organisms requiring a warmer climate migrated to lower latitudes, especially so during the coldest periods—those marked by periods of glacial buildup. During the warmer, interglacial stages, however, climates were warm enough for flora and fauna to migrate back to higher latitudes. This brings up the matter of global warming—a phenomenon well documented since the 1990s. The **cause** of global warming is not so certain; as some climatologists suggest, it may be a consequence of the fact that we are still receding from the last period of continental glaciation—a period which ended some 10,000 years ago. They note that the fossil record of some of the interglacial stages of the Pleistocene documents a climate warmer than even that of the present.

With Neogene fossils, maximum "strangeness" probably exists with its mammals, in part as a consequence of their large size. Also, a phenomenon **never** before seen in the fossil record appears in the Pliocene, the occurrence of **"worked flints."** Pliocene strata (in Africa) can yield crude stone tools. These tools are relatively rare and most are barely recognizable as having been artificially made but *these stone tools indicate a phenomenon new to the earth*, the appearance of a level of intelligence capable of interacting with available natural raw materials to assist in the business of living. We are, of course, looking at the **beginning of humanity**.

The Pleistocene Epoch (the ice age) finds the phenomena of tool making much more widespread so that at its end, some 10,000 years ago, stone tools become a significant part of the fossil record.

Marine Neogene Rocks

Coastal lowlands can be the location of limestone and marl deposited when either sea level was higher or (more likely) when these lowland areas were below sea level and later were uplifted. Parts of Florida as well as the Caribbean and Central America have Neogene marine strata which was deposited in this manner—and some of this strata can be full of fossil corals, mollusks, and less commonly, fossil echinoderms (echinoids). Overall, fossils of the late Cenozoic are some of the most diverse and accessible specimens of any part of geologic time.

Typical outcrop of late Cenozoic limestone in Florida and the Caribbean. This bluff, at the eastern end of the Dominican Republic, is Pliocene in age and contains fossil corals and mollusks.

Miocene limestone outcrop, Suwannee River, northern Florida. Florida has Cenozoic rocks covering the entire state; so all of its surface outcrops would (of course) be from the Cenozoic Era, with most outcrops of the southern part of the state consisting of limestone of Neogene age. This is a natural outcrop of typical late Cenozoic limestone.

Collecting Neogene Fossils

Late Cenozoic rocks, particularly those of marine origin, can be particularly productive in yielding fossils. Fossils in Cenozoic rocks are often more noticeable and obvious than are those found in older strata. Fossils in Cenozoic strata also can be well preserved, often being made of original material like shell or bone and not replaced or mineralized as is often the case with those found in older rocks. Late Cenozoic fossil occurrences sometimes are different from those of earlier geologic time, this especially being the case with those of the Pleistocene Epoch. Ice age mammals are especially unique in their occurrence compared to occurrences of earlier fossil mammals. Glacial activity and conditions that produced unique sediments like loess and glacial till are, in part, responsible for these unique fossil occurrences. These sediments, when they contain fossils, are different in the manner with which the fossils occur, both in preservation and in appearance, from those of older mammalian occurrences (and they generally are not mineralized). Different also are fossil occurrences in alluvial sediments of Pleistocene age, where there is a similarity with occurrences of modern bones found in soil.

Fine-grained Miocene ash-laden sediments: Fine grained sedimentary rock like that shown here can sometimes contain the imprints of leaves as well as yielding petrified logs, often well preserved ones. Burial of plants (and less commonly animals) by volcanic ash associated with explosive volcanic activity, like that associated with Mt. St. Helen's, can quickly bury life forms, preserving them with considerable fidelity. Note the two small faults in this road cut in eastern Washington State.

Miocene volcanic ash and gravel (conglomerate). Extensive Neogene volcanism occurred in the Pacific northwest of the U.S. and western Canada. Most of the late Cenozoic rocks and fossils of that region are associated with tuffaceous (volcanic ash bearing) sediments like this in Idaho.

Close-up of tuffaceous sandstone and gravel (conglomerate) found in the same area as in the previous picture, in eastern Washington State.

Shale beds exposed in a large ravine, Healy, Alaska. Most of these terrestrial sediments of late Oligocene to early Miocene age lack fossils; however, the red baked clay layer below the white beds crowning the outcrops top, contain large fossil leaves. Such red, baked clay is known as burnout. It formed from the spontaneous combustion of an underlying coal layer, which burned prior to the time when the canyon was carved.

Massive, silty sandstone of Miocene age exposed in the same canyon as in the previous photo near Healy, Alaska, 80 miles south of Fairbanks.

Colorful sequence of dipping (tilted) strata of Miocene age, which is tilted downward to the right (north). To the left (south) occur older Oligocene beds that tilt upward. Strata to the north (right) get younger as you go up the canyon. Note the red layer (burnout) in the sequence; it contains leaf impressions of subtropical Miocene trees. Strata at the head of the canyon, above the clump of trees, are Pliocene in age. Healy, Alaska, south of Fairbanks.

Tilted strata of late Miocene age with gravely layers at top left. These conglomerates are Pliocene in age, the black bands are thin coal seams inter-bedded with shale layers. Much of Alaska is geologically young and its younger strata, like this, can be tectonically deformed by being folded, tilted, and faulted. Generally, most late Cenozoic strata are not tectonically deformed like that seen here. Near Healy, Alaska.

Miocene coal beds cropping out along the Little Tonzona River, central Alaska. These beds, which form part of a thick coal seam (approximately 60+ feet thick), have been tilted by tectonic forces to a high angle. The coal beds lack any recognizable fossils, however burnout layers in this sequence to the west do yield numerous fossil leaves. Like most strata in Alaska, this coal bed has been tilted by tectonic activity since the material of the coal swamp was laid down some 15-18 million years ago—which geologically wasn't that long ago.

Late Cenozoic "rocks" cover a greater portion of the earth's surface than do those of most earlier parts of geologic time. This is especially true where large parts of the continents have a superficial cover made up of glacial deposits, till, loess or alluvial (river) deposits, all of which are more likely than not to be of Pleistocene age. This large area of (potential) outcrop gives greater possibility for the finding of fossils of this age then is the case with older strata. Like other fossil occurrences, however, Pleistocene fossils are usually localized in their occurrence.

Excavations and gullies eroded into loess can produce vertebrate fossils but these are generally rare, particularly considering the widespread occurrence of this wind blown sediment over large parts of the continents of the Northern Hemisphere. Late Cenozoic alluvial deposits also can be a source of vertebrate fossils, and here again, the Pleistocene—especially the late Pleistocene—can produce (at least locally) a variety of invertebrates and plants (usually cones and seeds and less commonly leaves), as well as vertebrates.

Phosphate mines near Lakeland, Florida, exposing Miocene through Pleistocene strata (Bone Valley and Hawthorn Formations). Numerous fossils have come from phosphate operations like this, which mine phosphate rock from the Miocene Bone Valley Formation.

Extensive phosphate pits exposing Miocene and Pliocene sediments near Lakeland, Florida. Photo taken in 1977.

Dragline digging phosphate rock of Neogene age, central Florida, 1977.

Columbia River Basalts. During the Miocene Epoch, huge amounts of mafic lava (basalt) were extruded from the earth's mantle to produce vast outpourings of igneous rock. Today these Miocene and Pliocene lavas (associated with inter-bedded sediments) cover a large portion of central Washington State and Oregon.

Large springs issuing from Pliocene Basalts, Thousand Springs, Idaho.

Snake River Canyon near Buhl, Idaho. Pliocene tuffs and tuffaceous fresh water shale were formed from extensive Neogene volcanic activity. In some areas such volcano-clastic sediment can locally contain vertebrate and invertebrate fossils.

Dipping leaf bearing Neogene tuffaceous shale (left) of the Miocene Latah Formation overlain by Pleistocene terrace deposits, eastern Washington State.

An example of a classic unconformity with tilted Cenozoic strata (Oligocene through Miocene) overlain by flat lying early Pleistocene conglomerate. Southern Idaho. *Photo courtesy of Warren Wagner.*

Pleistocene lake sediments from glacial lake Missoula, western Montana. Lake Missoula was a large lake that formed during the Pleistocene Epoch by the damming of a large stream system by ice. When this ice melted, a huge swell of water was released, which created giant ripples, producing what is known as the channeled scablands. Similar giant ripple marks which formed in a similar manner have been observed on the planet Mars.

Pleistocene sediments deposited in Lake Missoula, western Montana.

Mt. Rainier sticking above the clouds. The top of 12,200-foot Mt. Rainier, a strato volcano, has been built up from igneous activity taking place during the late Cenozoic. This means that Mt. Rainier is geologically young. This mountain, part of the Cascade Range, has taken at least three million years to form—a short span of time in geology.

Traversing a small stream in eastern Missouri by canoe to access Pleistocene sediments deposited in an ice age swamp.

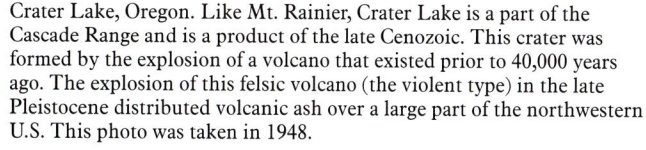

Crater Lake, Oregon. Like Mt. Rainier, Crater Lake is a part of the Cascade Range and is a product of the late Cenozoic. This crater was formed by the explosion of a volcano that existed prior to 40,000 years ago. The explosion of this felsic volcano (the violent type) in the late Pleistocene distributed volcanic ash over a large part of the northwestern U.S. This photo was taken in 1948.

Screening stream bottom sediments for Pleistocene vertebrate fossils in the previously shown stream—Eastern Missouri Society for Paleontology field trip, 2008. Pleistocene fossils found in streambeds like this are often a mix of warm and cold climate organisms. The warm climate animals represent interglacial stages and the cold climate ones are from periods of glaciation when entire ecosystems migrated southward.

A just-found mastodon leg bone plucked from a gravel bar of a small stream in eastern Missouri. Surface streams like this one, where the bone was collected, continuously cover and uncover bones and teeth of fossil and modern animals which, if not collected, would be moved or buried by the next flood. Arguments against collecting vertebrate fossils from surface streams like this generally are unjustified for this reason.

Members of Eastern Missouri Society for Paleontology looking for Pleistocene vertebrate fossils, eastern Missouri.

Many streams like this one expose Neogene sediments along their banks. Canoes offer an excellent method to access such outcrops where fossils contained in the beds are sometimes washed out and found in the streambed.

More creek bed Pleistocene fossil collecting.

This is a kid's fossil search for Pleistocene fossils in a small stream in eastern Missouri. Besides Pleistocene fossils, artifacts are sometimes found on gravel bars of such streams. Arguments **against** collecting artifacts from such streams are generally invalid for the same reason as stated above for fossils.

A vertical bank of loess: Loess is wind deposited silt. This loess was deposited during the mid-Pleistocene some 1.3 million years ago. Similar wind blown sediment occurs over a large portion of the central U.S. as well as in central Europe. Most loess in North America is of Pleistocene age, however Miocene and Pliocene loess occurs in China where this yellow sediment can contain well preserved vertebrate fossils. The Yellow River of China is named after the yellow sediment derived from its flowing through extensive beds of loess.

Travertine

A peculiar and distinctive rock type formed from spring deposits (sometimes hot springs {geothermal springs}) is known as travertine. Usually of Neogene age, travertine can sometimes contain the bones of animals as well as the shells of land snails, seeds or the impressions of leaves. Travertine is a type of limestone with a distinctive texture and it is frequently cut and polished where its interesting display of cavities is used as decorative stone in the interior of buildings.

Travertine is used as interior decorative stone. Here is the wall of a building lobby made of this distinctive hot spring-deposited limestone. This travertine came from quarries working the deposits shown in the previous picture near Cody, Wyoming.

Weathered travertine beds of Pleistocene age near Cody, Wyoming.

Travertine deposits of Pleistocene age: The water of Hot Springs, traversing through limestone beds, can dissolve limestone and redeposit it as a porous, distinctively textured rock known as travertine. Travertine sometimes contains the impressions of leaves, impressions of land snails or their actual shells, and more rarely the bones of vertebrates. These travertine deposits are in Yellowstone Park, Wyoming.

Sinkholes and Fissures

Neogene fossils are sometimes associated with solution phenomena where either fissures (enlarged joints in limestone) or related sinkholes acted as traps for animals or the bones of animals. Fissures, generally being elongate slots at the earth's surface, can effectively feed surface debris to lower levels of the fissure, which can include caves—bones found in caves often originating from associated fissures. Sinkholes also can act as mechanisms by which surface debris can feed deep into a sinkhole and then into a cave system. Both fissures and sinkholes can be prolific sources of bones, especially those of mammals of the late Pleistocene.

Junk filled fissure (sinkhole) in limestone of the northeastern Ozarks. Fissures like this have often been used by rural residents as a means of trash disposal. As surface soil feeds into the fissure, the trash also seems to disappear as it feeds into the fissure. In a similar manner the remains of animals can feed into a fissure over thousands of years and become associated with the sediments of underlying cave systems.

Caves

Neogene fossils can be associated with and found in caves, an environment conducive to preservation that is rare in earlier parts of geologic time. This is because most caves and caverns are geologically young—most of them having been formed during the Pliocene or Pleistocene epochs. Bones, tracks, and trackways of vertebrate animals are the types of fossils usually associated with cave deposits—most of them being from the Pleistocene Epoch and the majority of them from the late Pleistocene. Caves and their associated fossils also continue to be found as a consequence of construction, sinkhole collapse, and cave exploration. Virgin caves especially have the potential for yielding significant vertebrate fossils and, in some instances, newly discovered caves that have had no openings during the last few thousands of years can yield fossils older than the late Pleistocene, the age of most vertebrate fossils found in caves which have known natural openings.

Cave speleothems are composed of a secondary form of limestone similar to but usually more compact than travertine. Sometimes cave floors can contain the skeletons of Pleistocene (and Holocene) animals, which may be either partially or wholly covered by stalagmites like those on this cave floor.

The clay floor of this cave contains tracks and trackways, probably only a few thousands of years old. Passages like this in newly discovered caves can have tracks of considerably greater age as a cave entrance can become blocked for tens or hundreds of thousands of years, allowing any tracks, trackways or bones to be preserved. Cave floors like this or the clay of which the floor is composed can also contain the bones of fossil vertebrates. Unlike surface streams that can continuously cover and uncover fossil bones, bones in caves are not affected in this way and arguments against their indiscriminate collecting are usually valid.

Stalactites and columns (formed by the convergence of stalactites and stalagmites) are secondary forms of precipitated calcium carbonate. Similar to travertine, but generally formed by cooler water, such dripstone can cover walls and ceilings as they do in this Ozark cave.

The floor of this cave passage once accommodated a stream. When the stream bottom was wet, a raccoon walked on its clay surface and made these trackways. The stream since has cut some 30 feet into bedrock (dolomite) so these tracks are probably tens of thousands of years old—and hence are of late Pleistocene age. Generally most cave fossils like this date from the Pleistocene but a few can be older—depending upon the age of the cave system itself.

Here a large dripstone column rests upon a thick clay floor. It probably took at least 10,000 years to form this dripstone column so that this one started to form near the close of the ice age. At some more recent date, the clay floor on which the column rests compacted, causing the column to crack and settle.

Cave vertebrate fossils and collecting brings up the controversial issue of the appropriateness of the collecting of fossil vertebrates. This is because vertebrates, being multi-component fossils, are like a puzzle, and the more pieces of the puzzle that are available, the easier and more accurate will be the putting together of the final "picture." Cave vertebrate fossils, unlike those found in the beds of surface streams, are unlikely to be reburied or destroyed from weathering. The cave environment is a stable and un-varying one. For this reason, occurrences of vertebrate fossils in caves should usually be left alone. They often are protected and are considered a unique part of the cave environment, an environment that is also especially fascinating to a large number of persons.

An even more exotic medium of Neogene fossil preservation is found in asphalt or other petroleum seeps, the best known example being the Rancho La-Brea tar pits of Los Angeles, California. Similar occurrences, however, are found elsewhere over the globe, but this type of fossil occurrence is not common. Fossil resins can be another exotic preserving medium for Neogene fossils. Chapter nine is devoted to this, as well as to insects preserved in asphalt.

With Pre-Pleistocene fossils and sediments of terrestrial origin, terrace, and piedmont deposits can be a good source of fossil vertebrates and plants. In some parts of the world, upland gravels of Neogene age locally can contain the bones of animals, particularly those of mammals. Occurring in many parts of the high plains of the U.S. and Canada are sediments derived from material shed from the uplift of the Rocky Mountains. Sediments like the White River Series and the Ogallah Formation of the high plains represent examples of such deposits—deposits that locally can contain rich mammalian faunas.

Chena Hot Springs, Alaska: The author (left) enjoying a delight of geothermal energy in Alaska. Geothermal Springs are associated with those parts of the earth which are tectonically active and have had igneous intrusions (the source of the geothermal energy) emplaced within recent geologic time, that is since the beginning of the Neogene some 28 million years ago. The heat of geothermal springs like this is produced from the cooling of a mass of igneous rock below the surface—rock still in the process of cooling.

A slow moving stream in southern Florida. The bed of many streams in Florida as well as other rivers in the eastern U.S. can be a source of Neogene vertebrate fossils—especially mastodon and mammoth teeth, but also the teeth of the huge white shark *Carcharodon meglaodon*. These are sometimes recovered by diving 20-40 feet underwater to search the bed of the river.

Late Pleistocene or Holocene basaltic lava in southern New Mexico. Geologically young basalt composing flows like this will retain features like the pa-hoe-hoe flow structure seen here. If these lavas were of any geologic age at all, even Pliocene or Miocene, such features would have been eroded away. In a few rare cases fossil vertebrates and plants, including casts of the trunks of trees, can be found embedded in lavas like this.

Panoramic view of late Neogene (Holocene) basalt flows, southern New Mexico. Such lava came from the earth's mantle when North America moved westward over a spreading center of ascending, mafic magma—the same magma that produces mid-oceanic ridges.

Bluff of Miocene volcano-clastic sediments outcropping along the Yellowstone River in Yellowstone National Park. Yellowstone is a part of the earth's crust that has been influenced by a "hot spot," that is a region of upwelling molten rock originating from the earth's mantle. Periodic volcanic activity originating over the hot spot during the late Cenozoic has produced extensive amounts of volcanic sediments that sometimes preserve animal and plant remains as fossils. Thick sequences of coarse volcano-clastic sediments like this however generally lack fossils—they formed too rapidly and violently to bury things in any quantity to produce a fossil zone.

Late Cenozoic fossils, probably more than those of any other part of geologic time, are represented by the original material of the animal or plant. To some persons there is a particular appeal of this original-material-preservation. Often a cast of a fossil (be it manmade or natural) looks as real as, and effectively has most of the attributes of, the real thing; however, the real thing is usually more desirable. Why is this? One answer probably is that the actual fossil specimen (or the material of the original shell, bone or tooth) was around when the organism was still living. The material of the fossil has what can be looked upon as a more direct link to the geologic past. Perhaps this is one of the reasons that silicified fossils sometimes have less value with certain persons than do those preserved with original material. This value-of-the-original exists not only with fossils, but also occurs with artifacts as well as with a broad range of other antiquities. Late Neogene fossils also have ages that even the numerically challenged person can understand. Earlier materials, with ages measured in tens, hundreds or even thousands of millions of years, seem to illicit little interest with these same persons.

The author (left) and Virginia Stinchcomb collecting Pliocene fossils from a fresh road cut in Mississippi in 1955. Road cuts in soft Neogene sediments vegetate rapidly and become covered with grass or other vegetation. More recent highway and road construction now also cover such outcrops with sod so that they don't erode out into the numerous small gullies from which wash out fossils.

Value Range Used in This Book

Some persons have taken umbrage to the inclusion of monetary values on fossils in my works, considering it as being inappropriate, especially for someone like myself who has worked both as an educator and as a professional geologist. My rebuttal to this is that most fossils require a considerable amount of time and effort to collect. Good fossils are not that common. However, on the other hand, they do appear surreptitiously at times and often, if they are not collected, they either will be buried or destroyed. I have known people in the earth science professions who would not bother to pick up or collect a quality fossil (or mineral specimen) and I think this is sad—sad not only for the fact that if not collected most are lost but also sad for such a lack of personal interest. It's been said that a good collection of fossils really requires a certain amount of "blood, sweat, and tears" to acquire. If this seems a bit on the extreme side, it does emphasize that a dedicated desire is required to save, appreciate, and utilize such fossils. Also, the collection of a fossil is only part of the effort. Attractive specimens like those shown in this work also require preparation, a process consisting of the removal of rock or other sediment from the fossil. Some fossils also require various forms of cutting, polishing, and other laborious efforts to make them stand out. To consider that, after the expenditure of such a considerable amount of both effort and time, they should have no monetary value is ridiculous. Rather what should be considered is the dictum that the **scientific and educational value** of a fossil should take precedent over its **monetary value**. A person who accumulates a sizeable collection of fossils or who is in a position to acquire such will inevitably have some fossils that do have scientific value. This scientific value should take precedent over any monetary considerations. Sometimes however, it is only monetary considerations that allow for a fossil to be saved, as some persons working in the mining, excavating, and construction industries, where they might run into them, often have no interest in them *other* than a monetary one. The author would again like to state that educational and scientific value should take precedent over monetary considerations, but to negate monetary value is unrealistic, if for no other reason than without it many nice and important fossils would otherwise be lost.

Although seemingly (because of its title) more related to the Mesozoic Era than to the Cenozoic, an especially informative examination of current conflicts between academic paleontology and the fossil collecting and marketing community is found in the 2000 book *Tyrannosaurus Sue*—especially chapter four, "Taking a Howitzer to a Fly." Regarding legal complexities outlined in this book, Neogene strata is particularly culpable as vertebrate fossils are those draw the greatest number of laws and lawsuits because vertebrate fossils draw the most interest and therefore the largest crowd of anti-collecting hardliners in the field. The matter of legalities (especially letter of the law examples) regarding paleontologic collecting (to quote Richard Nixon) is a "real can of worms" and as a consequence of the "Sue" debacle, the situation appears to have recently become a lot worse. It's the authors opinion that if the effort and resources that have gone into trying to disenfranchise the collectors and dealers instead went into cooperation, real positive outcomes might take place—especially considering John Pojeta's statement, "The more eyes looking for fossils, the better for paleontology as more fossils will be found."

Rock, Mineral, and Fossil Fairs

A phenomenon that has increased considerably during the last few decades is the mineral and fossil fair or show. Throughout many parts of the globe, but especially in the U.S., shows featuring geo-collectables have been on the increase, while the traditional rock shop, which used to supply a similar clientele, has diminished. These rock, mineral (including meteorites), and fossil fairs can be looked upon as a type of temporary museum where participants can purchase and/or trade as well a view a plethora of specimens.

As conventional museums and/or science centers reduce or eliminate displays of geologic specimens in favor of technologically sophisticated exhibits with lots of bells and whistles, these temporary "museums" take on an even greater educational function. Considering this, it is, in the eyes of the author, sad that a few museum professionals and academics have attempted to discourage such fairs through support for import and export prohibitions and restrictions on both fossil and mineral specimens as well as encouraging other barriers to individuals' acquisitions of and "hands on" interactions with geo-collectables. It is hoped that this series of books, in a modest way, might create and solidify interest in the history of "mega time" through the "hands on" educational effectiveness of the geo-collectable.

Pricing of Fossils and Other Geo-collectables

Although the prime focus of a collection of fossils, minerals or other geo-collectable should be (and usually is) personal interest and the education of both the collector and others—many geo-collectables do have tangible monetary values. With the current (2009) instabilities of the monetary sector, such monetary value associated with quality fossils is perhaps reassuring. Note first the label "quality fossils"! Some fossils, like many bivalves and corals found in Neogene rocks, unless they are very attractive and well preserved, have little or no value. Late Cenozoic starfish, brittle stars, and vertebrates on the other hand do have monetary value, although it is the author's wish and belief that serious collectors place **interest** and **education** above **monetary considerations**.

Many other fossils, although rare, will be of interest to only a small coterie of collectors—rare late Cenozoic mollusks fall under this category. Others, such as mammoth and mastodon teeth, teeth of giant prehistoric sharks "meg teeth," complete fossil fish, and other vertebrates have an appeal to a greater number of persons and as such might be looked upon in terms of economic investments. One of the problems with fossils (as well as with other collectables), however, is their potential for liquidity if a collector needs to convert specimens to cash. Those types of specimens that are of interest to only a small group of collectors are especially vulnerable to a certain "lack of liquidity." This problem is so significant that some collectors limit their acquisitions only to those fossils in which there is broad interest.

MAPS Expo – 2009: A fossil fair held in Macomb, Illinois, each year is exclusively devoted to fossils. Most rock and fossils fairs include not only fossils but also a mix of minerals, meteorites, lapidary, and even artifacts. Note the Pliocene Stegomastodon at the left rear.

Doing this, however, limits ones exposure to the spectrum of what is actually out there and, as such, is a self imposed limitation and restriction on knowledge. The author is sympathetic to the real collector, but to the individual who "**high grades**" the fossil market, selecting only those "trophy specimens" to acquire as an assist to his or her ego or as a hedge against inflation, he has little sympathy.

Value Guide

With reference to the below value guide, a fossil with limited interest to most collectors may be given a value lower than are those to which there is a broader range of interest, even though the more desirable one may be more common than that specimen of limited interest. Pricing of fossils, as with most other collectables, generally is determined by what a seller is willing to take for the object and what the buyer is willing to pay.

Value range used in this book
- A $1,000-$1,500
- B $500-$1,000
- C $250-$500
- D $100-$250
- E $50-$100
- F $25-$50
- G $10-$25
- H $1.00-$10.00

Bibliography

Fiffer, Steve, 2000. *Tyrannosaurus Sue*. W. H. Freeman and Co., New York ISBN 0-7167-4017-6.

Pojeta, John, Jr. 1996. "Collecting Fossils" (abstract). Sixth North American Paleontological Convention, Washington D.C., Paleontological Society, Special Publication No. 8.

Triebold, Michael D., 1996. "Real World Solutions to America's Fossil Family Feud, a Personal Perspective." (abstract). Sixth North American Paleontological Convention, Washington D. C. Paleontological Society No. 8.

Wolberg, Donald L., 1996. "Laws, Regulations Policies, Conventions and Hazards in Paleontological Collecting, Buying and Selling." (abstract). Sixth North American Paleontological Convention.

Woodburne, Michael Q., 1996. "Fossil Collecting on Public Lands." (abstract). Sixth North American Paleontological Convention, Washington D. C., Paleontological Society Special Publication No. 8.

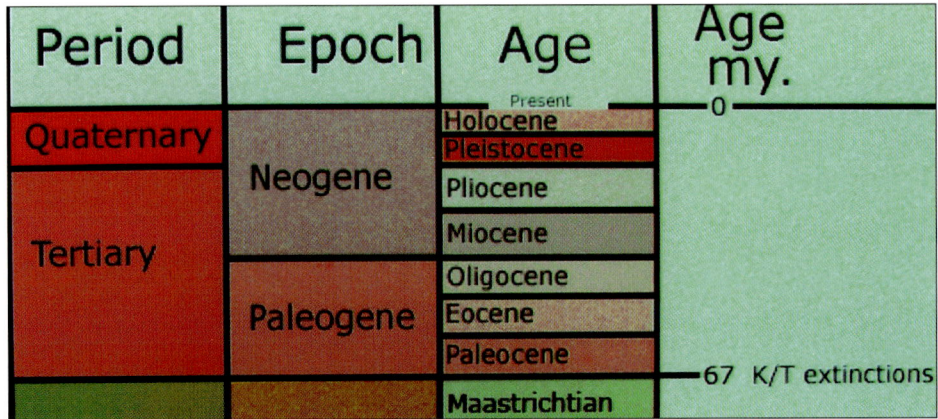

The Cenozoic Era (67 million years to the present): The Neogene begins approximately 28 million years ago with the Miocene Epoch and ends with the present.

TERTIARY PERIOD

Chapter Two
Plants

Stromatolites

Although they are **not plants**, stromatolites are produced by photosynthetic life forms—photosynthetic monerans known as cyanobacteria. Stromatolites represent the oldest direct evidence of life on earth, some occurring in rocks as old as 3.5 billion years.

Pre-Pleistocene Conifers

Conifers can be abundant fossil plants in late Cenozoic rocks. They include metasequoia, sequoia, and cypress of which the modern sequoia of California is a remnant from pre-Pleistocene time.

Stromatolite (transverse slice): Stromatolites are structures produced by primitive photosynthetic life forms (usually cyanobacteria, which are otherwise known as blue green algae). They are **not plants** but are placed under them as stromatolites are produced by a photosynthetic moneran (cyanobacteria). Stromatolites are still forming today in favored places and they are also the oldest known fossils. They provide evidence for life on the earth some 3.5 billion years ago, the age of the oldest stromatolites. These are geologically young stromatolites. Miocene, Svoge, Bulgaria. (Value range F).

Cypress: Latah Formation, Miocene, Spokane, Washington. (Value range F).

Stromatolite (lateral slice): Miocene, Svoge, Bulgaria.

Pinus destefani-side view: A lignitized pinecone. Pliocene. From a lignite mine south of Giovanni Valdarno, Toscana, Italy. (Value range F).

Pinus destefani: Top view of this flattened lignitized cone.

Metasequoia sp. in burnout from the previously shown locality in Alaska. As with Cenozoic coal beds in Canada (Alberta, B.C. and Saskatchewan) as well as in North and South Dakota, some Cenozoic coal beds of Alaska appear to have been derived from vast amounts of coniferous vegetation growing in a temperate or subtropical environment. (Value range G).

Outcrop of Miocene burnout near Farewell, Alaska, west of Mt. McKinley.

Pleistocene Conifers

Pleistocene fossil plants often include representatives of cold climates (from periods of glacial advance). They can also include plants of a more temperate climate, a climate similar to that which occurs today at the same latitude. Shown here are carbonized conifer cones (spruce cones) from a relatively cold climate, certainly one a lot colder than that which exists today at the latitude at which they were collected (39 degrees).

Spruce cones: Lignitized (or carbonized) cones from the previously shown outcrop. Spruce is a cold climate tree but during the ice age its range extended further south than it (naturally) does today. Pleistocene swamp deposits, eastern Missouri.

Spruce cone bearing layer, a zone exposed along a stream—often fossil localities are best reached by the use of small boats, like a canoe. Late Pleistocene swamp deposits, eastern Missouri.

Single carbonized spruce cone from the above locality in eastern Missouri. (Value range F).

Close-up of an outcrop of swamp deposits exposed along a stream. Note the abundance of fresh water snail shells.

Another view of the same specimen as above: Note also the abundance of fresh water gastropods.

Beech leaf: This partial leaf has been preserved in red burnout from the previously illustrated locality.

Angiosperms or Flowering Plants

Flowering plants (particularly fossil leaves) make up much of the obvious and collectable part of the Neogene fossil plant record. Shown here are some representative and large (or odd) angiosperm floras represented primarily by fossil leaves. The Late Cenozoic was primarily a world of angiosperms, like it is today.

Subtropical and Temperate Climate Flowering Plants from Alaska.

Sequences of late Cenozoic strata in Alaska occur that not only contain beds of low sulfur coal but locally can yield abundant leaf impressions of flowering plants that lived in subtropical or temperate climates. Often these are found in layers of red rock known as burnout. Burnout is formed from beds of clay occurring above coal seams that spontaneously caught fire. The combusting coal fires the overlying clay layers, often before the present topography was formed. Some layers of burnout contain excellent leaf impressions—sometimes large ones. Ages of these Alaskan Neogene plants range from Miocene to Pliocene.

Tilted or dipping thick coal beds which crop out along the Little Tonzona River near Farewell, Alaska. These thick coal beds, if they were in a region with roads and/or rail access would be a valuable energy resource. They are, however in a very remote region of the "Last Frontier" of the U.S.

Exposure of Miocene coal in strip-mining operation near Healy, Alaska. Thick, silty, and soft sandstone overlies the coal bed, which has to be removed before the coal can be strip-mined.

Dipping beds of Miocene strata. The red layers are leaf bearing burnout beds. The sequence of strata becomes younger to the right and strata exposed at the head of the small canyon are Pliocene or early Pleistocene in age. Note that all of these layers have been involved in tectonic activity, an indication of the recent geologic activity in Alaska. The Pliocene and early Pleistocene were just yesterday, geologically speaking.

Dragline removing overlying layers from Miocene coal, Healy, Alaska.

Colorful Miocene coal bearing terrestrial strata, Healy, Alaska.

Steeply dipping terrestrial strata of the Miocene age. Note the three black layers that are coal seams (or beds).

Talus of burnout containing poorly preserved leaves. Miocene, Healy, Alaska.

Betulia sp. Birch leaf: The lower beds of the Healy Creek Formation are considered to be Oligocene in age. These leaves, which come from a burnout layer higher in the formation, are from the Miocene Epoch. Healy, Alaska. (Value range F).

Group of Miocene leaves in burnout from outcrops shown in the previous photos, Healy, Alaska.

"Frilly" leaf. Healy, Alaska. (Value range F).

Group of leaves on slab of burnout. Healy, Alaska. (Value range F).

Maple seed, Miocene, Healy, Alaska. (Value range F).

Tuffaceous Sediments of the Pacific Northwest

Miocene and Pliocene angiosperm floras, represented by leaf compressions, are generally similar to the leaves of plants living today, more so than are those found in the older Paleogene strata, although the differences might not be discernible to anyone other than a paleobotanist who specializes in fossil angiosperms. On these pages are leaf compressions preserved in tuffaceous (volcanic-ash-rich) shale from strata that crops out and is sometimes inter-bedded with layers of lava in the states of Idaho, eastern Oregon, and Washington.

Group of leaf compressions from tuffaceous shales of the Pacific Northwest. Latah Formation, Miocene, Spokane, Washington.

Elmus sp. Elm leaf in tuff: Latah Formation, north of Boise, Idaho. (Value range F)

Outcrop of Miocene volcanic ash (tuff), northeast Oregon. Tuff beds in various parts of the Cenozoic of the Pacific Northwest can contain well (often beautifully) preserved fossils—the most abundant being fossil leaves and petrified wood. The latter is widely cut and polished by rock hounds or made into bookends. This wood often resembles modern wood like the trees that were killed and buried by the eruption of Mt. St. Helens. This wood was preserved in much the same way—being buried in volcanic ash, the silica of the volcanic ash replacing, on a molecular level, the original cellulose of the wood. This process is capable of preserving in great detain the original structure of the wood, even to the preservation of individual wood cells.

Acer sp. Maple leaf compressions in tuffaceous shale: A modern sugar maple leaf and its fossil leaf counterpart. Miocene strata north of Boise, Idaho. (Value range F).

Polished slice of petrified log from Miocene volcanic tuff, eastern Oregon: Petrified wood from Miocene volcano-clastic strata of Oregon and Washington is popular with rock hounds. It takes a high polish and looks like real wood. Petrified wood bookends are often made from this material, which is wood that was silicified (replaced by silica on a molecular level) after being buried by volcanic ash eruptions similar to that which buried trees in the 1982 eruption of Mt. St. Helen's. (Value range E).

Close-up of previously shown specimen: Note the maple seed at the left. Miocene volcanic tuff, north of Boise Idaho.

The Molasse of Switzerland and Southern Germany

The molasse is a thick sequence of sediment derived from the weathering and erosion of rocks that make up the Alps. As the Alps uplifted during the late Cenozoic, sediment produced by weathering, particularly clay, silt, and sand, was washed northward, filling in a lowland area between the Alps and the Jura Mountains (Schwabisch Albs) of Bavaria. The mass of sediment deposited between these two ranges is known as molasse—it can locally be rich in fossil plants as well as in fresh water invertebrates and fossil mammals. The term molasse comes from the Latin word for soft (same root as in mollusk and mollify). Rocks of the molasse, like many of the Cenozoic, are soft compared to the older hard rocks, which were the source material for the sediments that make up the molasse.

> The Miocene formations of Switzerland have been called *Molasse*, a term derived from the French *mol*, and applied to a *soft*, incoherent, greenish sandstone, occupying the country between the Alps and the Jura. This molasse comprises three divisions, of which the middle one is marine, and being closely related by its shells to the faluns of Touraine, may be classed as Upper Miocene. The two others are freshwater, the upper of which may be also grouped with the faluns, while the lower must be referred to the Lower Miocene, as defined in the next chapter.

Explanation of the molasse from Charles Lyell's *Elements of Geology*, 1854.

A group of angiosperm leaves preserved in a clay lens from the upper layers of the molasse.

Populus balsamoides: A popular leaf in clay from the middle Miocene portion of the Molasse. (Value range F).

35

Populus latior: Another species of popular, Gunzburg, Germany. (Value range F).

Liquidambar sp. A sweet gum leaf. (Value range F).

Ulmus pyramidalis. (Value range F).

Salix sp. A willow, Gunzburg, Germany. (Value range G).

Acer tricuspidatum. Maple: If it looks like a duck and quacks like a duck it probably is a duck—a saying that can be applied to Neogene leaves as well. These look like maple leaves and no doubt they are, however with the older Paleogene leaves, what looks like the leaf of a living tree may actually be from quite a different type. Pollen of many of the trees living today are absent from beds yielding Paleogene leaves, so that *some* paleobotanists believe that the modern-looking, older leaves, may not really be closely related to trees of today. With the Late Cenozoic a leaf is probably what it looks like—in this case, a maple leaf. (Value range F).

Cinnamomum polymorphum. Cinnamon is a tropical plant. The climate of central Europe, as well as that of central North America, was tropical in the Miocene. It would become cooler in the Pliocene and then culminate in the ice age of the Pleistocene. (Value range G).

Daphnogene polymorpha.
Gunzburg, Germany.
(Value range G).

Daphnogene polymorpha. (Value range G).

Polished petrified log section. Petrified wood can occur with the leaf bearing beds of the molasse. (Value range F).

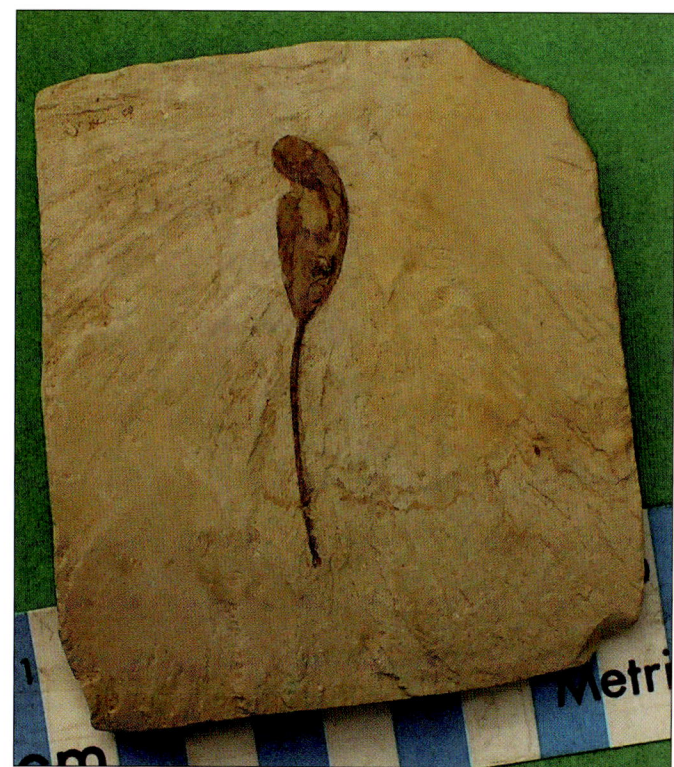

Gleditsia knorrii. (Value range G).

Booklet (in German) on the Molasse fossils: Numerous books like this one, aimed at serious fossil collectors, are published in Europe on significant fossil bearing strata like the Miocene molasse.

A few other Neogene fossil plants and a modern one.

Reed or pond grass stem from sediments of the Steinheim astroblem. During the Miocene, a small asteroid (or large meteoroid) hit what is now southern Germany. This produced a crater in Jurassic limestones, which then became a lake. Limy sediments on the lakes bottom preserved a variety of fossils, including plants like this. This is a type of grass, a monocot—the grasses having first appeared at the beginning of the Miocene became a dominant type of vegetal cover on the earth since that time. It was the appearance of grasses that allowed browsing animals, like horses, to evolve and proliferate.

Quercus sp. (Oak): A species of oak from Miocene diatomite of the West Coast of the U.S. (Oregon). The lignitized leaf contrasts with the white diatomite in which the leaf is embedded. Troup Creek, Oregon. (Value range F).

Quercus sp. Oak: A sinkhole in Paleozoic limestone (Knox Formation) of eastern Tennessee filled with sediment during the Miocene Epoch and preserved a variety of life forms, including plants and vertebrates. Known as the Gray Fossil Site, excavation of this occurrence is an on going project that has produced early Neogene fossils from a region where previously none were known. Leaves of oaks are the most common fossils at the Gray site and they have entered the fossil market in some quantity. (Value range F).

Sabal Palm: Impression of a large pale leaf in ferruginous (iron bearing) conglomerate. Pliocene? "Lafayette Formation," Northern Mississippi? An example of a serendipitous fossil find in Neogene conglomerate beds. *Courtesy Dept. of earth and Planetary Sciences, Washington University, St. Louis.* (Value range: unique and rare)

A group of Miocene Grey Fossil Site leaves at MAPS Expo. The working of this filled sinkhole deposit (paleokarst) is an ongoing project. A museum has been emplaced at the site and the sale of common fossils from the site like these oak leaves goes to help finance the excavation which, like many filled sinks or paleokarst fossil occurrences, is unique. This arrangement of placing commonly occurring fossils from a unique deposit onto the fossil market is a win-win situation for all involved, in contrast to an exclusive-monopoly attitude taken at some paleontological sites.

Impression of charred wood in basalt: Flows of fluid, black lava (basalt), sometimes can engulf trees and logs, the charred wood leaving impressions like this in the basalt. Igneous rocks (other than volcanic tuff) rarely contain fossils, but charred wood impressions are an exception. Wood impressions like this are found in the basalt flows of Hawaii, one of the few fossils found in that state. This specimen is from Pliocene or Pleistocene lavas of New Mexico. *Courtesy Dept. of earth and Planetary Sciences, Washington University, St. Louis.*

Most fossil plants found in Cenozoic strata are from trees, be they leaves, fruit, seeds or pollen (a type of microfossil not covered in this work). Herbaceous plants, so common today, have left a scant fossil record, but there is evidence that many of them, especially those often referred to as "weeds," are of very recent geologic origin. Other modern plants, like the giant yuccas shown here, grow in places where erosion rather than sedimentation occurs (uplands) and thus are rarely preserved as fossils. Plants that grow in such arid regions, like cactus and these yuccas, are not in an environment where they might be buried by sediments and thus preserved as fossils—as a consequence, the fossil record of arid uplands is almost entirely unknown. (These giant yuccas formed a backdrop for WWI {The Great War} soldiers while training at Camp Grant, in western Texas, in 1917). *Photo taken by Carl Schlueter, 1917.*

Basalt field (Craters of the Moon National Monument, Idaho): Basalts here, like those of New Mexico, range in age from Pliocene to recent.

Chapter Three
Miocene and Pliocene Corals and Mollusks

Corals

Coral can be common fossils in Neogene limestones. Late Cenozoic limestones were often formed at the edge of the continents when shallow seas covered them, an environment often ideal for the growth of corals. Later, these sediments were exposed by uplift—ending up above sea level and then subjected to erosion and weathering with the fossil coral weathering out in full detail. Cenozoic corals belong to an order known as hexacorals, a coral order which first appeared in the Mesozoic Era (mid-Triassic). All corals living today are hexacorals. They are characterized by having six or a multiple of six septa. (Septa are vertical plates that are part of the coral's skeleton.) Other fossils commonly found in late Cenozoic limestone are the shells of mollusks. In this chapter are shown representative Neogene molluscan fossils as well as some that were replaced with silica: a type of fossil popular with rock hounds.

Outcrop of Pliocene limestone, 45 km west of Santo Domingo, Dominican Republic:
This is a typical outcrop of late Cenozoic limestone in the Caribbean and Central America.

Tampa Bay coral, Pliocene: These fossil corals have been replaced with a finely crystalline form of quartz known as chalcedony; they are especially popular with rockhounds. (Value range F).

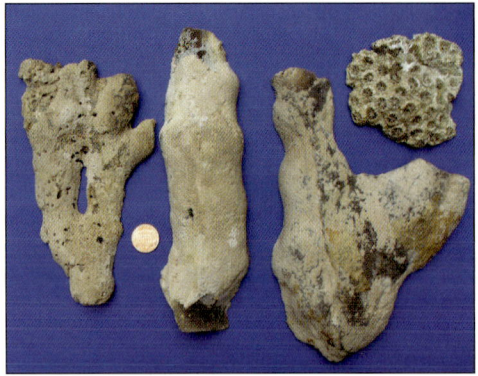

Tampa Bay coral exteriors: The exterior of these silicified corals faintly show the small corallites (the structure that housed an individual coral animal). Tampa Bay fossil corals are generally more interesting in their interiors, with a bubbly (botryoidal) structure of chalcedony, than their coral surfaces—their exteriors usually being drab and vague.

Silicified Miocene colonial coral—its calcite composition replaced with chalcedony. This specimen has been made into a bookend. Tamiami Formation, southern Florida. (Value range F).

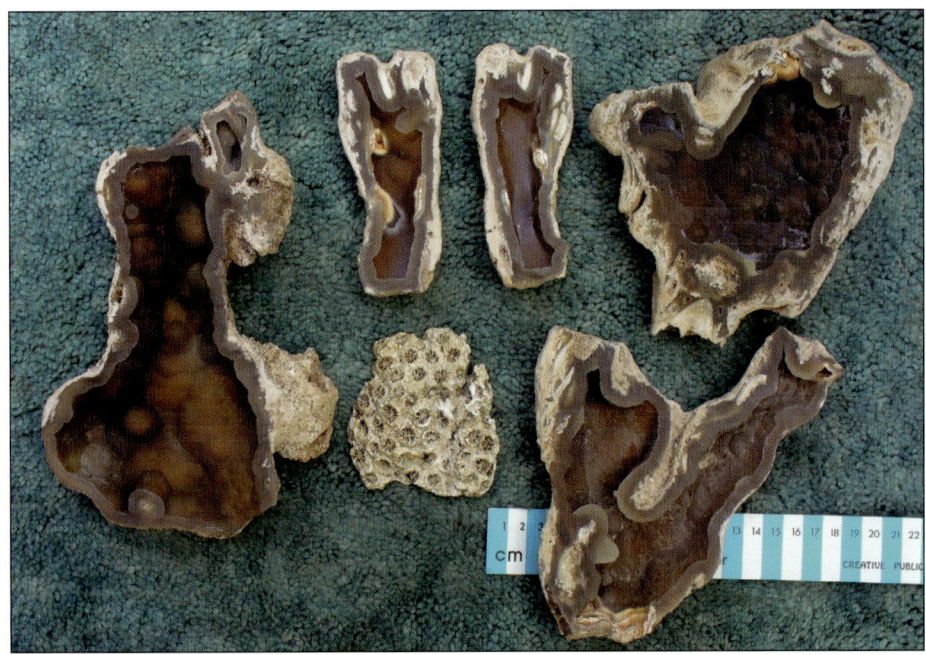

Tampa Bay coral geode interiors: The interiors of these Pliocene corals are lined with chalcedony, a cryptocrystalline variety of quartz. The coral head acted as a nucleus for its replacement by chalcedony. Tampa Bay geodized corals are as much minerals as they are fossil specimens and have been collected by and popular with rock hounds for years. (Value range F).

Interior of a silicified Neogene coral: These attractive geodized corals are as much mineral specimens as they are fossils. The peculiar interior pattern shown is known as botryoidal structure and is a phenomena associated with the process of silica replacement of the coral head.

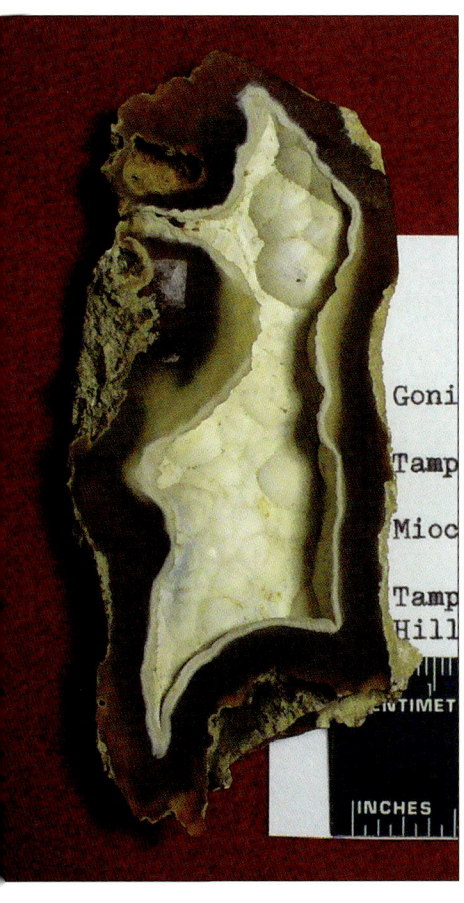

A single Tampa Bay coral geode with its botryoidal chalcedony interior. Tampa Bay Formation, Pliocene.

Outcrop of Pliocene Limestone in the eastern Dominican Republic where the following fossils were collected. Here excavations for a water main uncovered fossiliferous coral and mollusk bearing limestone. Temporary exposures of rock like this may serendipitously yield nice, and sometimes scientifically significant, fossils. Excavations like this in Florida, Central America, and the islands of the Caribbean can, at times, yield similar fossils.

Fossiliferous Pliocene limestone fragments produced from working outcrops shown in the previous photo. White, pure Neogene limestone like this is characteristic of the Caribbean and southern Florida.

Cabachons made from silica (quartz) replaced Neogene hexacorals. Silicified coral can be colorful and rockhounds work such corals into interesting (and attractive) semi precious stones like these. These specimens come from Thailand. (Value range G, single specimen).

Casts of hexacorals: Pliocene and other Neogene corals are known as hexacorals because they have a multiple of six radiating septa (septa are the radial structures preserved in this coral cast). Fossil corals can be common fossils in Late Cenozoic rock of southern Florida as well as in Neogene limestone beds of the Caribbean where these originated. The radiating septa stand out particularly well with these coral molds. (Value range F).

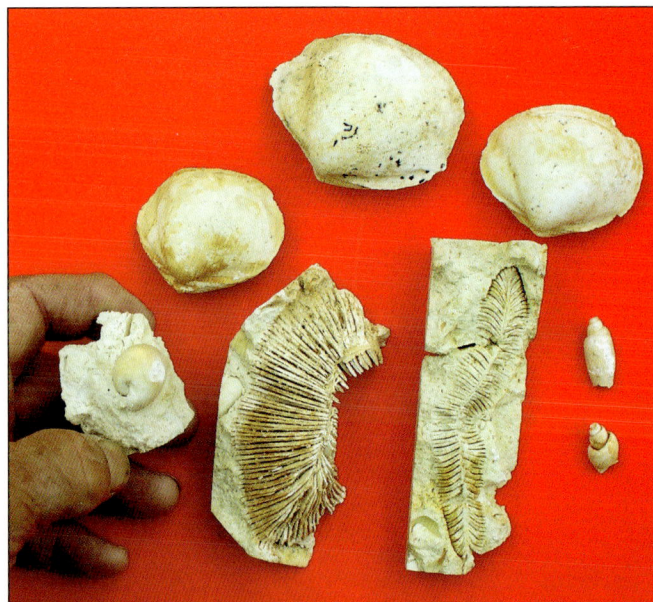

A group of natural invertebrate fossil molds from Pliocene limestone, which includes fossil corals, eastern Dominican Republic.

Hexacoral—Pliocene, Jamaica: Corals like these are found weathered from limestones which crop out along the coastal parts of Jamaica as well as in other parts of the Caribbean where late Cenozoic limestone can form high sea cliffs and outcrops. (Value range F).

Mollusks-Pelecypods

Neogene marine fossils resemble (somewhat) those of the earlier Paleogene, however, they are closer to (or the same as) those living in modern oceans. Indeed, the original designation of the Miocene—the earliest epoch of the Neogene—was established by Charles Lyell as the epoch having the majority of its molluscan genera (70%) still living. This applies to many marine organisms of the Miocene, but was especially applicable to its mollusks—fossils that are some of the most common to be found in marine strata of this age. Late Cenozoic marine limestone can be particularly fossiliferous. Some of this limestone, upon close examination, is almost entirely composed of fossils. Typical examples of this limestone are those seen in southern Florida, as well as in parts of the Caribbean. This rock, often pure white or cream colored, can be full of fossils. Extensive Neogene fossil faunas also occur in Europe, particularly in France and Italy. Indeed, Neogene marine limestone and marls occur in many places where relatively low lying coastal regions, once submerged under marine waters sometime during the past 28 million years, were later uplifted—these regions now exposing fossil bearing Neogene rock representative of the Miocene, Pliocene, Pleistocene, and Holocene epochs of the Cenozoic Era. Neogene marine fossils can be some of the most abundant and common fossils, sometimes resembling the bleached shells of clams and snails (or coral) found on a beach today and which might be ignored by fossil collectors for that reason.

Pectin coalingensis: Pliocene, San Joaquin Formation, Kettleman Hills, California. The genus Pectin is an abundant clam genus (pelecypod). It has existed over a major part of geologic time (since the Devonian Period). Pectins in the seafood market are known as scallops. This is a fossil scallop shell from Miocene strata of the West Coast; it came from the Kettleman Hills of central California, a locality that has produced large numbers of well-preserved Pliocene fossils. (Value range G).

Lycopecten estrellanus (Conrad). A fossil oyster: The shells of oysters can be common fossils occurring in marls deposited at the margin of continents or islands like those of the Caribbean. This specimen is from southern California. Santa Margarita Formation, Miocene. (Value range G).

Pelecypod—internal and external mold: This interior and exterior mold of a marine clam occurs in white, Pliocene limestone typical of the Neogene. Montego Bay, western Jamaica. (Value range F).

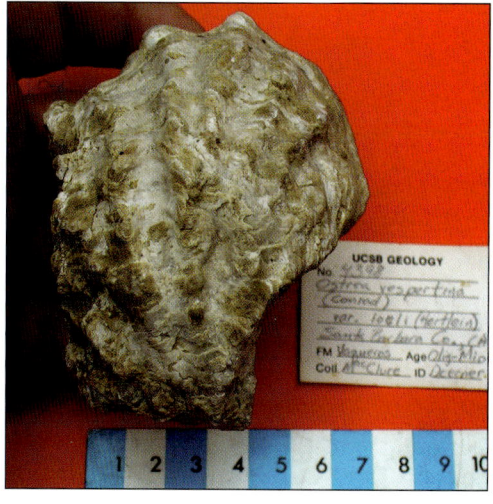

Ostrea yespertina (Conrad). A fossil oyster: The shells of oysters can be common fossils found in marl that was deposited at the margin of continents. Vaqueros Formation, Miocene, Santa Barbara Co., California. (Value range H).

Ostrea titanensis: A very large fossil oyster from a well-known and well-collected locality in central California. Pliocene, Kettleman Hills, California. (Value range E).

Pectin (Lyropecten) estrellanus Conrad: A pectin found in tuffaceous mudstone which forms sea cliffs in parts of southern California. Some of the rocks that form both the coastal ranges and sea cliffs of California were "plastered" onto North America as the continent has moved westward and has overridden the Pacific plate. These impressions of Pectins (scallops) have come from such a zone. Miocene, Vaqueros Formation, southern California. (Value range E).

Mollusks, Gastropods

Gastropods of the Neogene are very similar to those living today.

Ecphora quadricostata Say: A distinctive and collectable gastropod of the Pliocene Epoch. Yorktown Formation, Lee Creek Mine, Aurora, Beaufort Co., North Carolina.

Calliostoma philanthropum (**top snail**).

Another group of the distinctive gastropod *Ecphora quadricostata*.

Buccinum sp.: Pliocene. Southern California.

Crucibulum lawrencei: This spatula-like gastropod (limpet) is characteristic of the late Cenozoic. Yorktown Formation, Lee Creek Mine, Pliocene, North Carolina. (Value range G).

Vasum loddini: Pliocene. Tamiami Formation, Pinecrest Member, Pliocene. Sarasota shell pit, Sarasota Florida. (Value range G).

Vasum loddini: Another view of this ornate gastropod.

Buccinofusus parilis: Miocene, St. Mary's Formation, St. Mary's River, Maryland. (Value range G).

Murex globosus Emmons: Modern murex shells are collectable gastropod shells. The body of this mollusk was the source of the purple dye known to the ancients as **royal purple**. Tamiami Formation, Pinecrest Member, Sarasota shell pit, Sarasota, Florida. (Value range G).

Vermicularia sp.: These are peculiar "worm-like" gastropods. Pliocene, Lee Creek Mine, Yorktown Formation, Aurora, Beaufort Co., North Carolina. (Value range F).

Vermicularia sp.: Pliocene, Lee Creek Mine, Yorktown Formation, Aurora, Beaufort Co., North Carolina. (Value range G).

Campanile defrenatum: Avesa, Verona, Italy. These internal molds of huge snails come from Miocene limestone beds quarried near Verona, Italy. (Value range E)

Crepidula princeps Conrad: Pliocene, south of Ventura, California, Ventura Co., California. A peculiar gastropod containing a unique calcareous pocket (brood chamber). (Value range G).

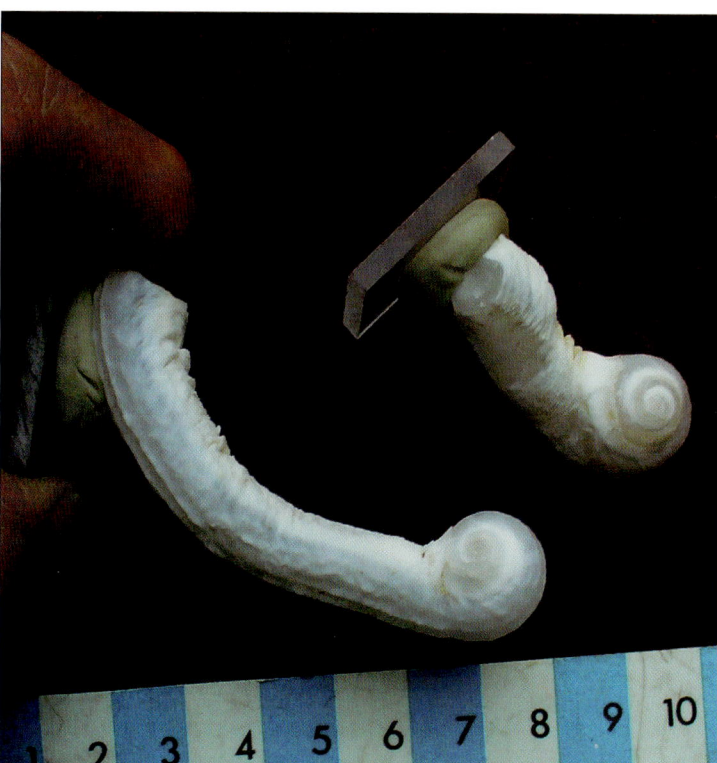

These peculiar silicified (quartz replaced, specifically with chalcedony) gastropods are from the Pliocene of Java, Indonesia. (Value range F).

Pelecypods from the Pliocene of the Paris basin, Touraine, France. (Value range H).

Another fat snail, Touraine, Paris Basin, France. (Value range G).

Group of Pliocene gastropods, Touraine, Paris Basin, France. (Value range F).

Turritella sp.: Pliocene, Touraine, France. (Value range G).

Fat snail: These well preserved mollusks occur in soft, calcareous sandstone. Pliocene, Touraine, Paris Basin, France. (Value range G).

Mollusks, Cephalopod

Argonauta: Thin shelled brood chambers of a Miocene **octopus**. Argonauta somewhat resemble the shell of an ammonite; this has led some paleontologists to suggest that the shell-less octopus is a descendent of this Mesozoic life form. Cephalopods generally are rare fossils in Neogene marine strata, especially when compared with their abundance in older marine strata. The living octopus has a shelled brood chamber exactly like this Miocene specimen. Miocene, Marecchia River Formation, Messinian Stage, Rimini, Italy. (Value range F, single specimen).

Chapter Four
Arthropods and Echinoderms

Arthropods – Barnacles

The arthropods and echinoderms of the Neogene generally are similar to those living today.

Barnacles are arthropods, although they don't look like it! Unlike most arthropods, adult barnacles live attached to hard items on the sea floor and are stationary. They generally live in shallow waters of the continental shelf. Fossil barnacles can be locally common in Neogene rocks that formed at the margins of continents, usually occurring in marls. Those barnacles found in late Cenozoic limestones and marls generally are the same as living species or are close to living species.

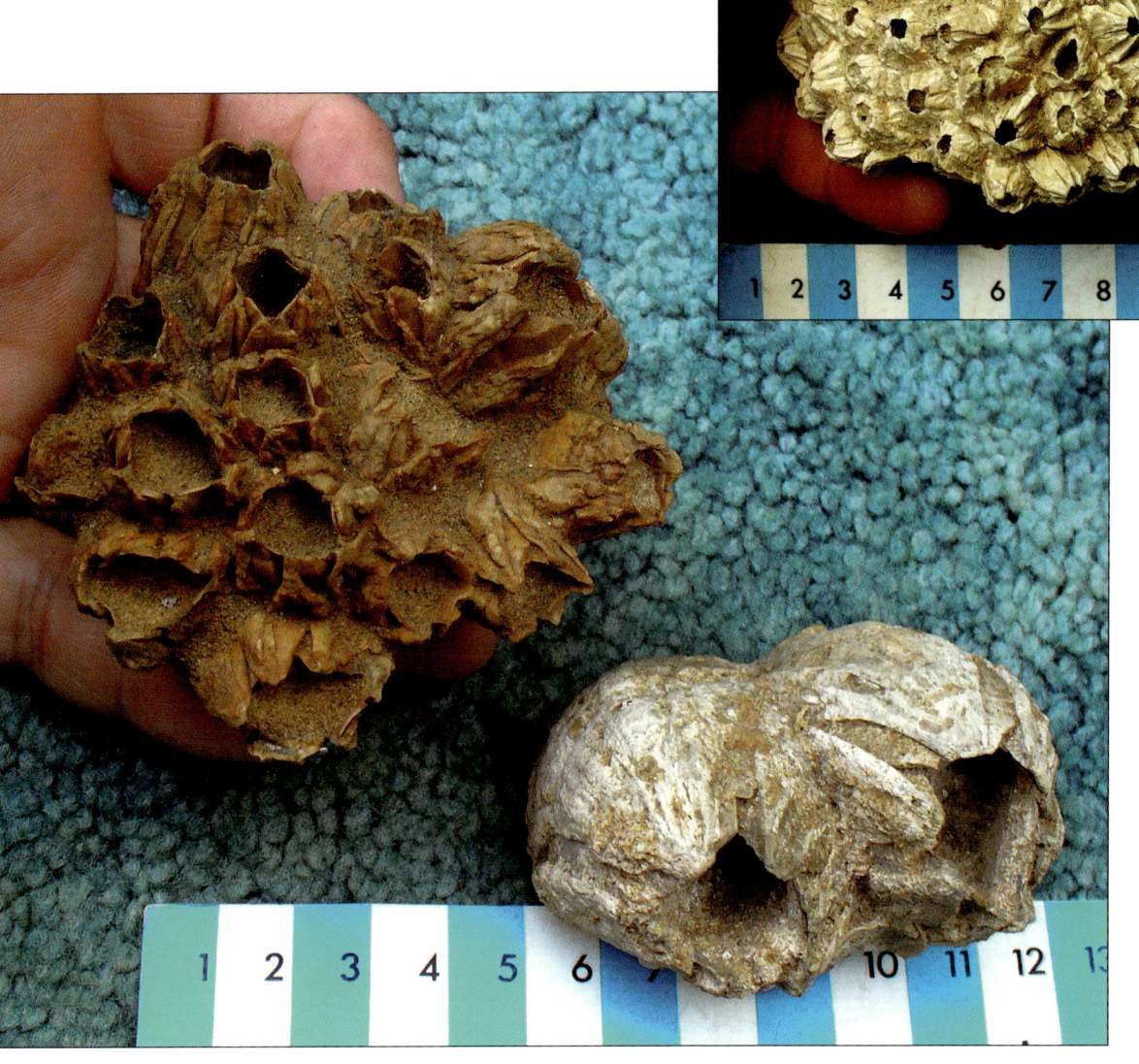

Balanus sp.: A single specimen of a barnacle. Pliocene, Kettleman Hills, Kings County, California. (Value range G).

Balanus sp.: A typical colony of barnacles attached to part of a fossil shell. Pliocene, Kettleman Hills. Kings County, Southern California. (Value range G).

Arthropods – Crabs

Fossil crabs often occur in concretions which, when broken open with a hammer or by the freeze-thaw cycle, can sometimes reveal a complete animal inside with pincers like those shown here. Neogene crabs are morphologically (and presumable genetically) similar to those living today.

Leptomithrax (Trichopeltarion) greggi. Another group of New Zealand spider crabs. (Value range F).

Leptomithrax (Trichopeltarion) greggi: A spider crab in a nodule. These nodules erode from shale beds that accumulate in either streambeds or can concentrate in the talus beneath sea cliffs where wave action wears the concretions into water worn cobbles. Miocene, Canterbury, New Zealand. (Value range F).

Callinectes sp. Blue crab, (male): The crab genus *Callinectes* is widespread in modern oceans. Miocene, St. Mary's Formation, St. Mary's River, Maryland. Specimen found as a stream pebble extricated from a concretion by the freeze-thaw cycle. (Value range F).

Leptomithrax (Trichopeltarion) greggi. A group of four of these spider crabs in water-worn concretions. The bottom specimen has the iron stain of the mineral limonite. Miocene, Canterbury, New Zealand. (Value range F, single concretion—part and counterpart).

Pinnixa galliheri. A group of three specimens of pea crabs (pinnotherid crab) from a (probable) deep-sea environment. The crabs occur in strata of the Monterey Formation, part of which is a deep water, hard cherty sequence whose tilted strata can form the sea cliffs of southern and central California. Monterey Formation, Carmel Valley, central California. (Value range F, single specimen).

Crab of the genus *Cancer* sp. preserved in a concretion. These crabs are found in Pleistocene marls, which outcrop in the vicinity of Galveston, Texas. They occur in concretions that wash out of terrace deposits exposed along the Texas Gulf Coast. The crab bearing concretions occur in Pleistocene marls, which constitute Pleistocene Terrace deposits, Port Lavaca, Calhoun County, near Galveston, Texas. (Value range E).

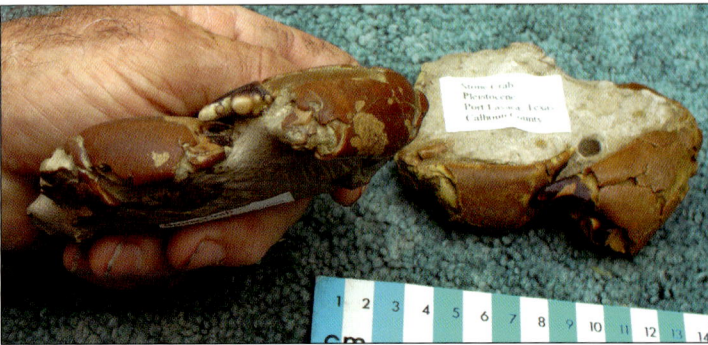

Cancer sp.: These crabs are found in hard concretions, which weather from marls making up Pleistocene terrace deposits occurring along the Texas Gulf Coast. Note that some of the original coloration (black pigment) is preserved on the crab's pincers. Port Lavaca near Galveston Texas, Calhoun County, Texas. (Value range E).

Pinnixa galliheri: This small crab in siliceous, deep-sea sediments is one of the fossil crabs shown here which has not been preserved in a concretion. Monterey Formation, Carmel Valley, central California. (Value range F).

Carcinus exculptus: These crabs come from clay beds that also yield spectacular (and unusual) fossil fish (*see Chapter Five*). Miocene (Messinian), Marecchia River Formation, Poggio Berni, Rimini, Northern Italy. (Value range F).

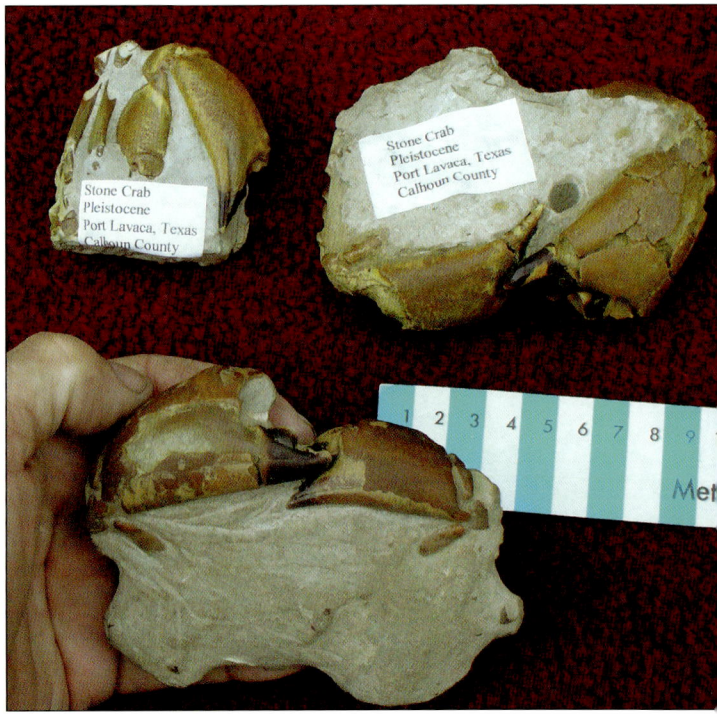

This is a group of nice stone crabs. Note original coloration of the pincer tips. These crabs are preserved in hard, sandy calcareous concretions which weather from Pleistocene terrace marls. Port Lavaca, Calhoun County, Texas. (Value E, single specimen).

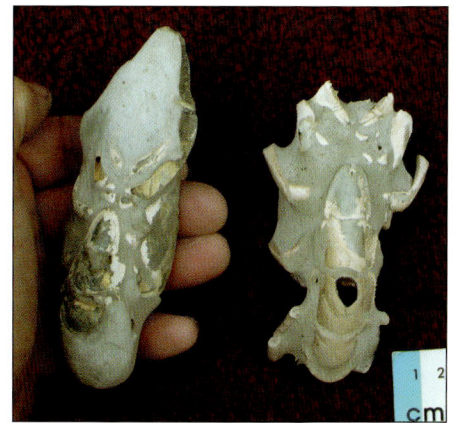

Thalassina anomala: Two of these small lobsters are preserved in light colored concretions, Daly River, Darwin area, northern Australia. (Value range F, single specimen).

Cancer sp.: Another view of the group of stone crabs of the previous photo.

Arthropods – Lobsters and Crayfish

Lobsters and crayfish are obviously related to each other, lobsters being marine crustaceans and crayfish being smaller crustaceans associated with fresh or brackish water. Both animals, when found as fossils, often occur in concretions that formed around the body of the crustacean. Both of these arthropods make desirable and collectable fossils.

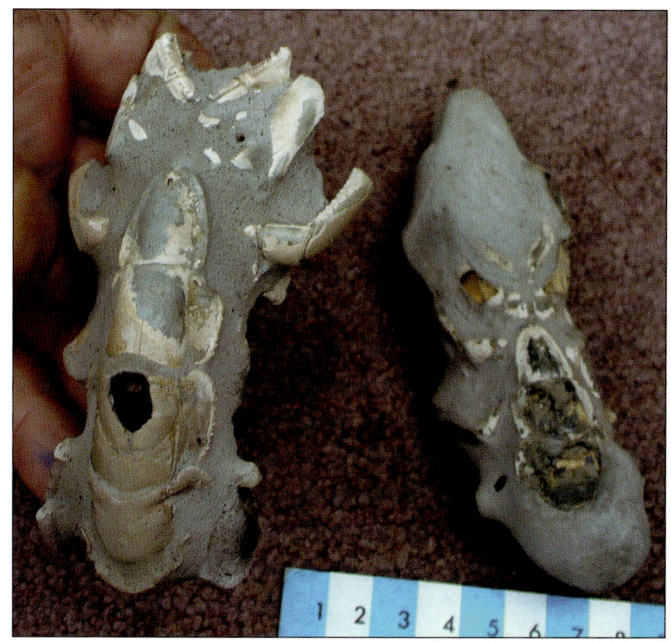

Thalassina anomala: Another look at two of these small mud lobsters. Daly River, Darwin area, northern Australia.

Thalassina anomala: A single, dark specimen of a small lobster that has been preserved in a concretion. These fossils are found in quantity in the Darwin area of northern Australia near the mouth of the Daly River. It is typical for crustaceans to be preserved in concretions like this. The associated concretion is hard. but these fossils are geologically quite young. Lobsters of the genus *Thalassina* are still found in the area where these fossils were collected. Some believe that these "fossils" may be only a few thousands of years old and are from the Holocene rather than the Pleistocene Epoch, even though they are in hard, stony concretions. If Holocene, technically they would not be fossils at all as they would not be old enough. The general "rule of thumb" being that to be a bona fide fossil, it has to be at least Pleistocene in age and the Pleistocene Epoch ended 10,000 years ago. (Value range F).

Arthropods – Insect Larva and Small Crustaceans

Larva of water breeding insects locally can be preserved in sediments deposited on the floor of ancient lakes in which the insect larva lived. Shown here are insect larvae preserved as compressions in freshwater limestone, which formed from sediment deposited in ancient lakes. Such fossils, preserved in fresh water or lacustrian (lake deposited) limestone, can be quite collectable.

"Fresh water crayfish:" Ancient lakebeds, formed in a lake that occupied the valley of the Snake River of southern Idaho, were the source of these fossils. The lake was formed in Pliocene time by the damming of the ancestral Snake River by lava flows. This lake existed into the Pleistocene Epoch when sediments that accumulated in it were eroded by the modern Snake River producing a deep valley, the walls of which (in part) are composed of these Pliocene lakebed sediments. The area where these crayfish were collected is now off limits to collecting, having become the Hagerman Fossil Beds National Monument. Glenn's Ferry Formation, Late Pliocene, Gooding County, Idaho.

Captoclava longipoda. Different specimen preserved as a compression in a more carbonaceous (dark colored) rock from the same locality as specimen shown above. A number of nice specimens from fresh water lake limestone came onto the fossil market through the Tucson show in 1998. Lingu County, Shandong Province. China. (Value range F).

Crayfish in nodule or concretion: Pliocene Snake River Valley Lake deposits. Glenns Ferry Formation, Gooding County, Idaho.

Captoclava longipoda. Different specimen preserved as a compression in a more carbonaceous (dark colored) rock from the same locality as specimen shown above. A number of nice specimens from fresh water lake limestone came onto the fossil market through the Tucson show in 1998. Lingu County, Shandong Province. China. (Value range F).

Captoclava longipoda. The two specimens shown in the previous photos under different lighting conditions. Miocene, Lai-Yang County, Shandong Province, China.

Silicified small crustacean: Small, calcareous nodules from Miocene strata near Barstow, California, contain these silicified arthropods which can be extracted by dissolving the nodules in weak acid.

Libellula doris. Larva of a darter dragonfly. This larva lived in a Miocene lake whose sediments buried and preserved dragonfly larva that had not metamorphosed into dragonflies but instead fell to the lake bottom and were buried by lake sediment, which preserves them. Santa Vittoria d'Alba, Cuneo, Italy. (Value range F).

Silicified small crustacean from Miocene lakebeds near Barstow, California. (Value range F).

Libellula doris. Group of dragonfly larva from the same Miocene occurrence as above. Santa Vittoria d'Alba, Cuneo, Italy.

Insects Preserved as Inclusions in Amber

Amber is a fossil resin exuded from resin generating trees. The resin is then buried and preserved in sedimentary rock, rock usually of brackish or fresh water origin. Insects and other small organisms become stuck in the exuding resin when it is liquid and sticky, the resin then becoming hard and over time undergoing a polymerization process, forming the hard, clear material known as amber. The best-known amber occurrence is that of the Baltic region of northern Europe surrounding the North Sea. The second is in the Dominican Republic, where amber bearing strata occur in the mountains at the eastern end of the island. These amber bearing zones of Hispaniola are mined with hand tools by locals—the amber then being polished, graded, and those pieces containing insects and other fossil inclusions being set aside to sell at a premium price. The specimens shown here were purchased in a local shop in the Dominican Republic at the eastern end of the island. Dominican amber ranges from Oligocene through Miocene in age. Pleistocene copalite is another fossil resin also obtained on the island. Copalite is softer than amber and specimens from the Dominican Republic have not yielded insect inclusions like those found in the older Dominican amber.

Determining the geologic age of various fossil resins is often difficult as they are usually found in strata almost devoid of other fossils—particularly the fossils of marine mollusks which form the basis for the determination of the various epochs of the Cenozoic Era. Also, fossil resins sometimes are found from an earlier epoch to have worked their way into more recent epochs via natural processes (migrating by burrowing animals, mixing during excavations, eroding out of one depositional layer and tumbling into another, etc.)—the individual amber pieces having been derived from different sources—this appears to be the case with some of the Dominican amber.

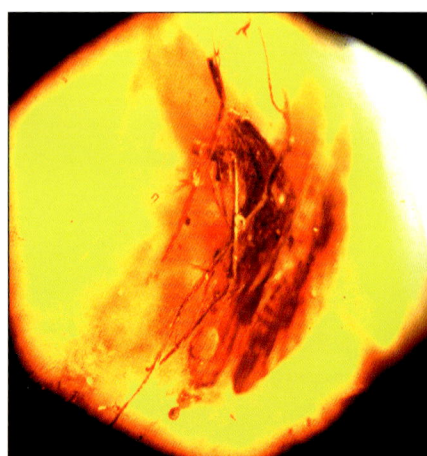

Beetle in Dominican amber, Miocene. (Value range F).

Flies in Dominican amber. Miocene. (Value range F).

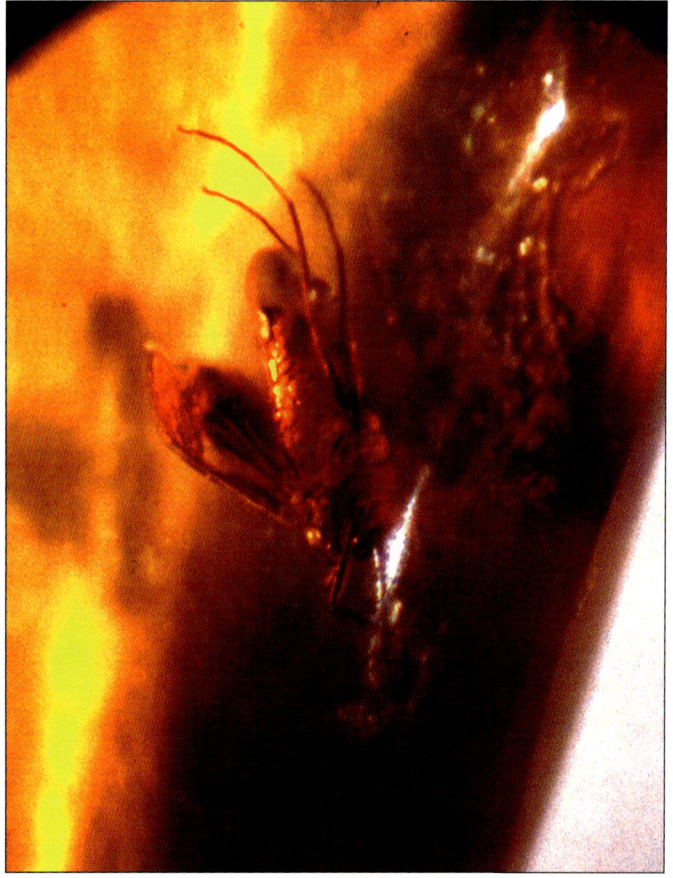

Fly in Dominican amber.

Fossil Insect Nests

Shown here are fossil mud dauber nests preserved in Pleistocene fossil soil zones (paleosoils) of Australia. Similar fossil mud dauber nests occur in Oligocene fossil soils of the bad lands of South Dakota.

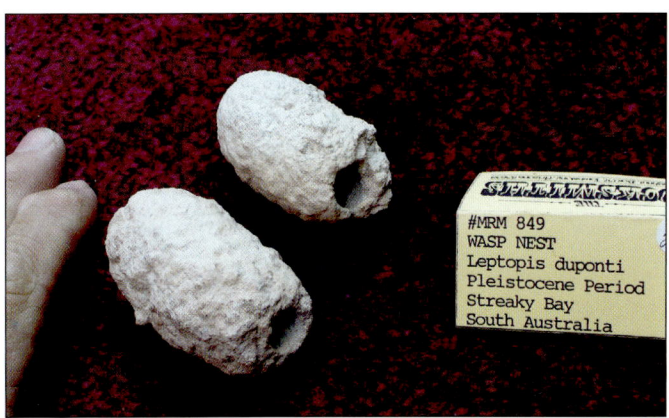

Leptopsis duponti: Wasp or mud dauber nest. Pleistocene soil zone, Streaky Bay, South Australia. Similar fossil mud dauber nests are found in ancient soil zones (paleosoils) of Oligocene age in the White River Group of the Bad Lands of South Dakota and Nebraska. (Value range F).

Tripneustes parkinsoni: Round, regular echinoids with an irregular urchin like *Eupatagus* in the center. The species name, *parkinsoni*, is named after Dr. Ellis Parkinson, a nineteenth century physician-paleontologist better known for medically recognizing as a specific disease what previously was known as the shaking palsy and is now known as Parkinson's Disease. Miocene, Burdigalian, Vaucluse France.

Miocene and Pliocene Echinoderms

Echinoids, or sea urchins, are frequently found in limestones deposited during the Neogene. Miocene echinoderm species are the ones most likely to be those that are extinct. About half of Pliocene species are still living and the majority of Pleistocene echinoderm species are still living. Other than echinoids, echinoderms such as crinoids, asteroids (starfish) and holothurians (sea cucumbers) are rare fossils in Neogene strata. When these fossils do occur, they can represent echinoderm "population explosions" where a population of starfish or serpent stars ballooned into large numbers and then crashed. Modern asteroid populations, like the crown of thorns starfish, have been found to do the same thing in today's oceans.

Mellita sp.: A lovely fossil sand dollar with a lovely name. Pliocene, Kettleman Hills, California. (Value range F).

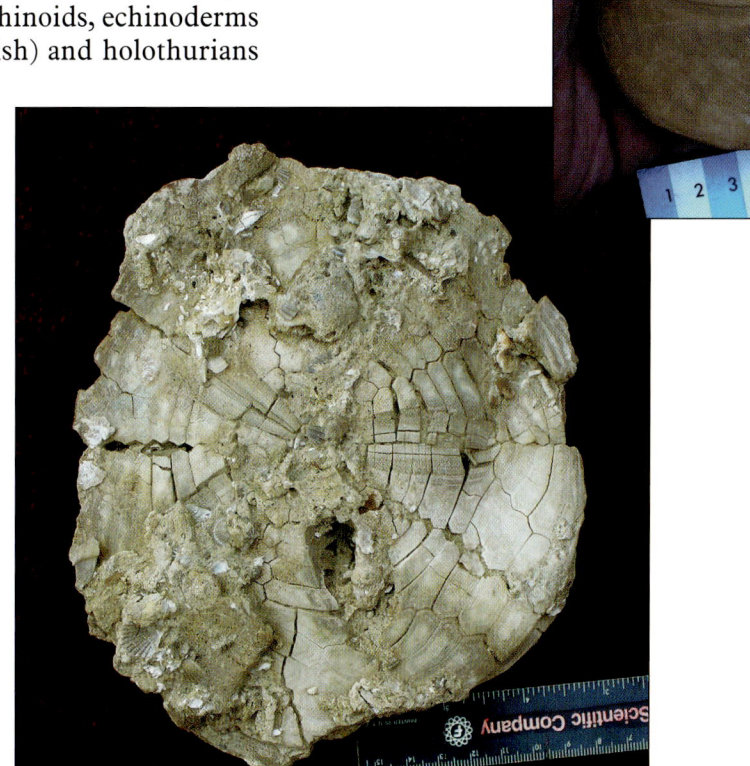

Scutella albertella Conrad: An aberrant echinoid suggestive (in some way) of earlier types of echinoderms. Its plates are pronounced, as is the case with Paleozoic echinoids. Miocene, St. Mary's River, St. Mary's County, Maryland. (Value range F).

Dendraster coalingensis: A common and well-collected sand dollar from this well-known locality in central California. Upper Miocene, Kettleman Hills, California.

Ophiocrossota baconi: (Population explosion of ophuroids). Echinoderms sometimes undergo localized population explosions where large numbers of them overwhelm everything else living in the area. An example of a modern occurrence of this phenomenon is the spread of the "Crown of Thorns" starfish onto coral reefs, which are adversely affected by them. This slab is part of a Miocene population explosion of serpent stars (ophuroids), which form an ophuroid limestone layer in the Santa Margarita Formation. Santa Margarita Formation, Miocene, Arroyo Grande Canyon, San Louis Obispo County, California. (Value range E).

Clypeaster brevior Sequenza: A large, peculiar hump-shaped echinoid. Miocene, Helvetian Sand, Tropea, Italy. (Value range F).

Ophuroid limestone: A concentration of small brittle or serpent stars from a "population explosion" of these small echinoderms. Zones of these echinoderms occur sporadically in Miocene and Pliocene strata at the western edge of North America. This slab is from Baja California, Mexico. Tirabuzón Formation, Corkscrew Hill, North of Santa Rosalia, northern Baja California Sur, Mexico. (Value range E).

Late Pliocene brittle star bed—flip side of the same slab shown previously: The 2 cm thick slab is entirely composed of these small ophuroids. Tirabuzón Formation, Corkscrew Hill, north of Santa Rosalia, northern Baja California Sur, Mexico.

Bibliography

Burns, Jasper, 1991. *Fossil Collecting in the Mid Atlantic States*. The Johns Hopkins University Press. Baltimore and London. ISBN-8018-4145-3.

Grimaldi, David A., 1996. *Amber—Window to the Past*. The American Museum of Natural History. ISBN-88109-1966-4.

Chapter Five
Sharks, Rays, and Fish

Sharks

Neogene sharks are similar in many ways to the sharks of modern oceans, with one notable exception. This exception is the existence of gigantic sharks like *Carcharodon megalodon*—a great white shark. This "great white" was a shark as big as that in the movie *Jaws*; but unlike the Hollywood creation, it actually lived—it flourished during the Miocene and Pliocene epochs.

Small sharks' teeth: Moroccan phosphate deposits.

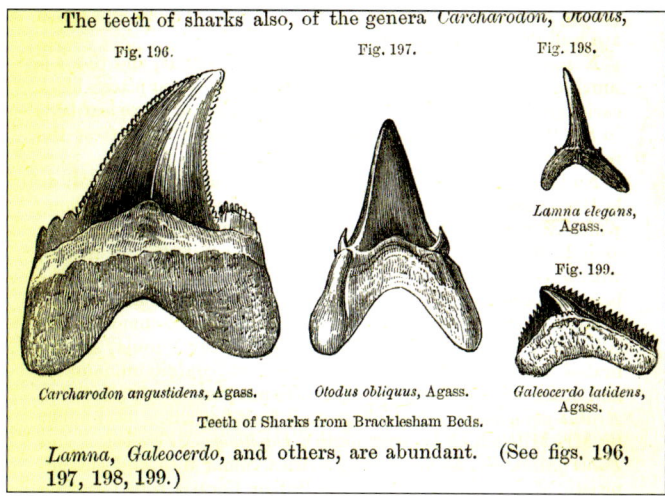

Explanation and illustration of Cenozoic fossil sharks' teeth from Lyell's *Elements of Geology*.

Odontaspis sp. and other small sharks' teeth: Phosphate mines in southwestern Morocco produce a considerable variety of vertebrate fossils. The most common examples are sharks' teeth that can occur in large numbers. These shark and ray teeth have been widely distributed among collectors and dealers at rock shows and are particularly appealing to children. (Value range F for entire group).

Jaw of modern White shark: This is one of the largest sharks living today. The great white shark, *Carcharodon megalodon,* was 4-5 times larger than this and lived during the Miocene and Pliocene Epochs.

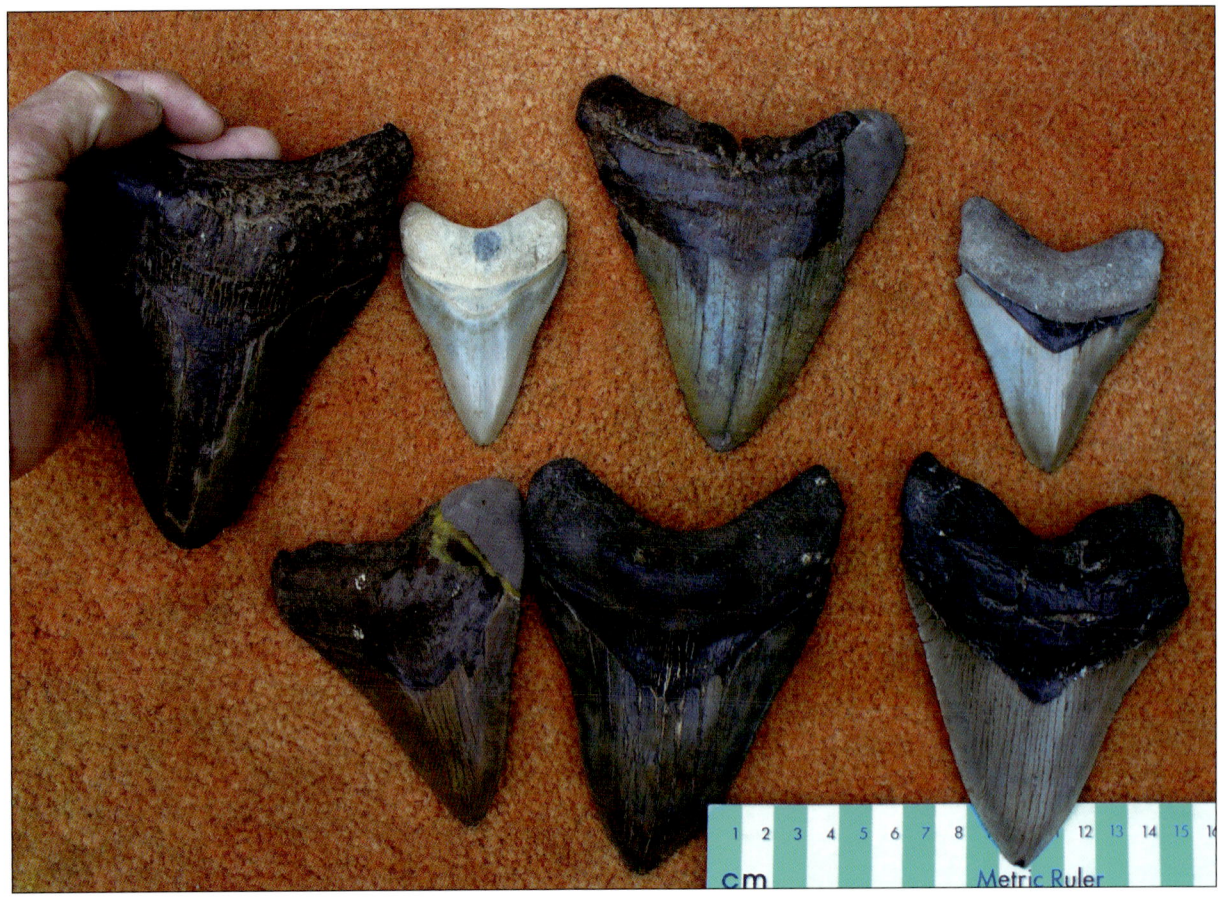

Carcharodon (Procarcharodon) megalodon: These large teeth are from an **extinct great white shark**. They come from river outcrops where the Cooper River of South Carolina cuts into Miocene strata. These gigantic sharks lived in the Miocene and Pliocene epochs—their large teeth are quite popular with collectors and indeed fascinate the public at large. A considerable number of fine *Carcharodon* teeth (meg teeth) have come from the Cooper River, as well as from other stream bottom sediments of the Miocene age on the East Coast of the U.S. (Value range E, single tooth; value range B for group).

Carcharodon (Procarcharodon) megalodon: Another group of these formidable teeth of this great white shark from Miocene outcrops of South Carolina. (Value range E for group)

Carcharodon (Procarcharodon) megalodon: This large shark's tooth came from sediments on the floor of the Pacific Ocean. Trawling the deep ocean floor has yielded not only these teeth, but also the bones of whales–some of which look quite fresh but are a few million years old, as are these teeth. Many of these deep-sea fossils show an etched surface, as waters of the open ocean, unlike those of shallower regions, are not saturated in calcium. Under this environment they can dissolve, as the deep-sea environment will remove calcium from them—this is the reason that calcareous microfossils like foraminifera are absent in deep-sea sediments. This etching process is particularly noticeable at that part of the tooth where porous bone is present. Bone is more subject to solution and etching than is the much harder tooth enamel. These teeth are believed to be Pliocene in age—the Miocene and Pliocene epochs being the time of the giant **megalodon** white shark (Value range F, single tooth).

A reconstructed jaw of *C. megalodon* made with megalodon (meg teeth) from Miocene strata of South Carolina.

Group of *Carcharodon* teeth: These fresh looking teeth of the great white shark came from deep sea sediments of the Pacific Ocean.

Carcharodon megalodon: Close-up of teeth from sediments on the floor of the Pacific Ocean. Note erosion of bone on that part of the tooth not covered with porcelaneous dentine enamel. These fossil teeth appear to be a first—the first fossils on the fossil market to come from the 60% of the earth's surface that is covered by deep oceans.

A group of Megalodon teeth from ocean floor sediments of the South Pacific at a MAPS expo. Darker specimens to the right are from Miocene strata of the eastern U.S.

Outcrop of Miocene sediments—phosphate pit, central Florida: Many high quality fossils come from the Miocene Bone Valley Formation, which is exposed in pits like this. Occurring with the Miocene fossils are those of Pliocene and Pleistocene age, which occur above the Miocene layers mined for phosphate rock. Also abundant in these excavations are fossil stingray spines and parrot fish teeth.

Rays

Teeth and the barbed tail of sting rays can be common fossils in Neogene rocks. They are particularly common fossils in late Cenozoic rocks of Florida.

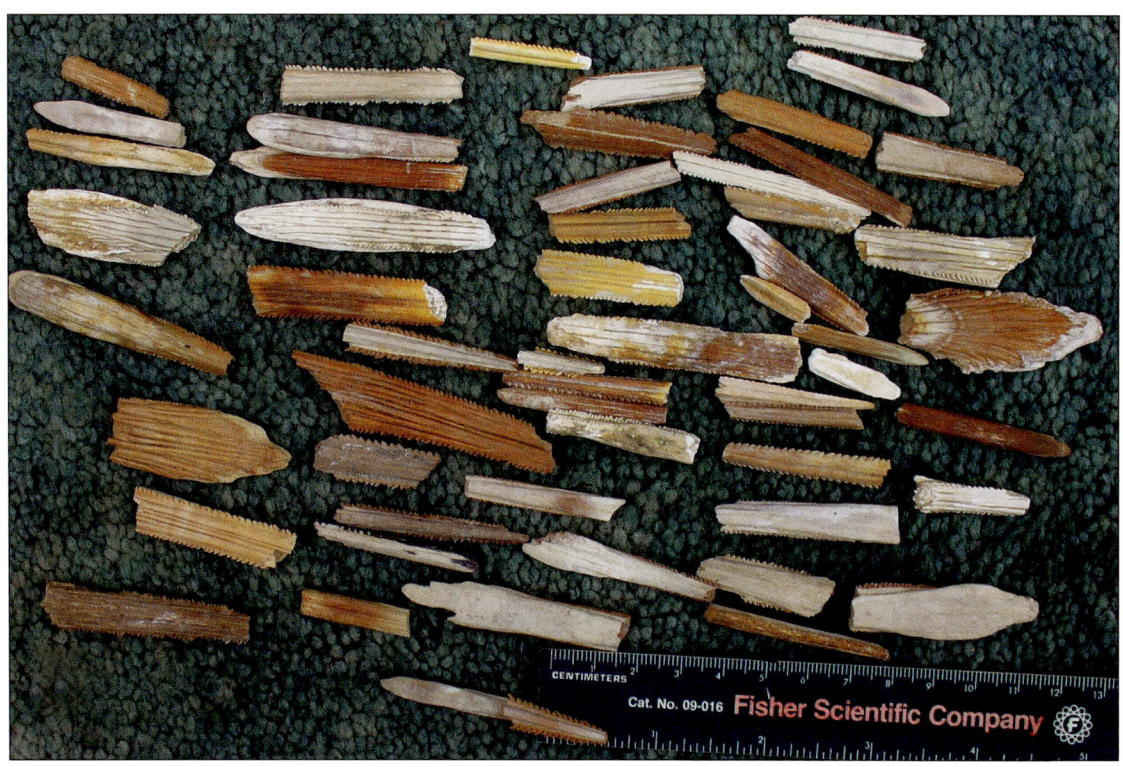

Ray spines from phosphate pits dug into the Miocene Bone Valley Formation of Florida.

The spines of stingrays can be common fossils in Miocene sediments of many regions of Florida as well as elsewhere where late Cenozoic marine strata occur. Bone Valley Formation, Miocene, Lakeland County, Florida.

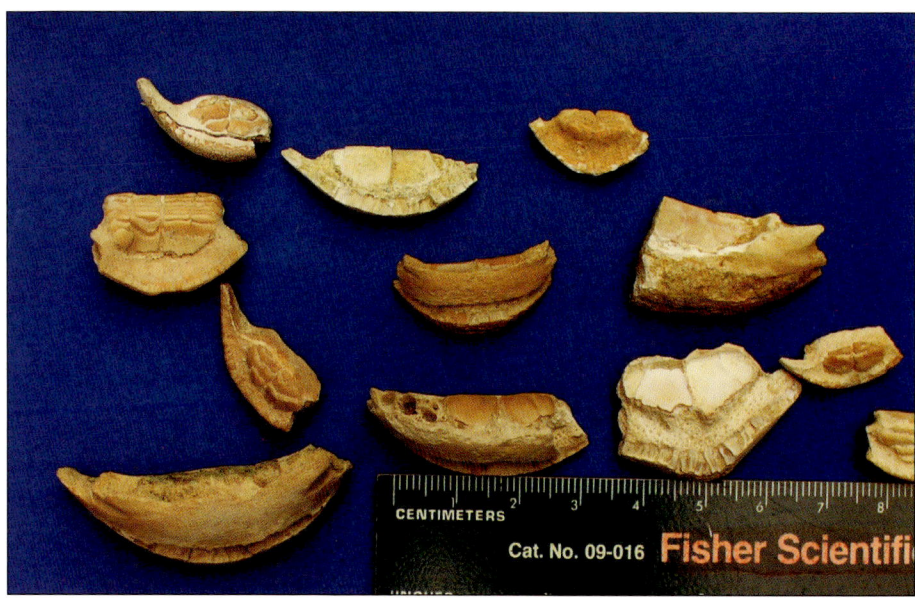

Diodon sp.: Porcupine or parrotfish teeth, Miocene, Lakeland, Florida. Parrotfish have hard, firm beaks with which they browse on coral. These beaks can be common fossils in Miocene and Pliocene marine sediments like those of the Bone Valley and Pungo River Formations. They are usually found associated with stingray spines. (Value range F for group).

Tinca. sp. (Tench): Fossil fish from lake deposits formed in a lake produced by an asteroid impact crater now known as the Steinheim impact structure (or basin). Meteorite or asteroid impact at the earth's surface can produce a crater which then can become a lake, as happened during the Miocene Epoch in what is now southern Germany. (Value range F).

The crater, shortly after formation, filled with fresh water and became a lake in which a variety of organisms lived. The lake then gradually filled with limy mud eroded from surrounding brecciated (broken) impacted Jurassic limestone—rock which was particularly susceptible to solution as it was highly fractured from the asteroid impact. These lake sediments offered an ideal medium for preserving the remains of plants and animals living in the lake which, when buried on the lake bottom, became well-preserved fossils on the lake sediment bedding planes.

Steinheim Basin Fossils

The group of fossils shown here were once parts of living organisms which lived in a lake produced by the impact of a small asteroid. Some twenty-five million years ago, during the Miocene, a relatively small asteroid (compared to the size of the asteroid which produced the K/T extinctions) impacted the region of southern Germany, now the location of the town of Steinheim, and formed a medium-sized impact crater. The crater itself has been partially eroded away and filled in with sediment—such a "has been" crater is known as an **astroblem**.

Here is a geologic map of a terrestrial impact structure (Crooked Creek astroblem), a structure similar to the Steinheim structure. Detailed geologic mapping of areas of anomalous geology can map out circular structures, which, after much geologic debate, are found to delineate ancient impact sites. Such an impact site, because of terrestrial erosion and weathering over geologic time, no longer exists as a crater, but can still be identified from its brecciated rock and its other peculiar geologic features. This type of geologic structure produced from extraterrestrial impact is known as an astroblem.

Tinca sp.: Another fossil fish from sediments deposited in the impact-formed lake of the Steinheim Basin. (Value range F).

Booklets from Steinheim Basin Museum: These booklets discuss the occurrence of this astroblem and its associated fossils. Geologic features in Europe often form the basis for local museums, which highlight a particular geologic attraction, such as the Steinheim Basin. These two publications discuss (in German) the Steinheim Astroblem and its associated fossils.

Shatter cones: High velocity impact on a planetary surface produces these distinctively shaped cone-like structures with radiating striations. This group of shatter cones came from the Steinheim Astroblem (or basin). It is formed from hard Jurassic limestone where its cone-like shape and distinctive striations indicate phenomena that, along with some high-pressure minerals, are a "hallmark" of extraterrestrial impact.

Breccia from an astroblem: The impact of an extraterrestrial object (asteroid) formed this breccia (broken, angular rock fragments cemented together). This breccia came from the Crooked Creek Structure of southern Missouri; similar breccias occur associated with all astroblems. Such a breccia is formed from the intense breakage of rock on high velocity impact. Impact produces not only a crater, but also a large amount of rock fragments as well as rock flour. The physical condition of this impacted material allows it to dissolve and go into solution—in the case of the Steinheim Astroblem, such brecciated rock was limestone, which partially dissolved and precipitated again to form the limy sediment deposited at the bottom of the Miocene Lake that occupied the crater.

A group of fossil fish preserved in limy siltstone deposited on the bottom of the impact formed lake—the impact that produced the Steinheim structure. (Value range E for group).

Breccia from the Steinheim Basin: High velocity impact capable of producing a crater on the earth's surface can also produce other distinctive impact phenomena like this monomict breccia. This breccia is made up of angular fragments of intensely fractured Jurassic limestone. Impact by the asteroid that formed the Steinheim crater severely fractured the rock, forming a breccia. Along with the breccia 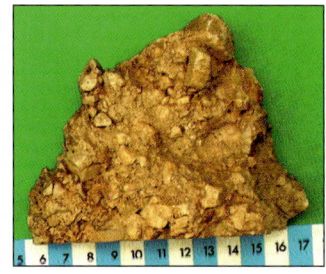 was formed rock flour (highly pulverized rock), which, because of its large surface area, readily dissolved and on precipitating, cemented the breccia together. Breccias, formed in a similar manner (except for the solution part), are found on the surface of the Moon and in many meteorites (which usually come from the Asteroid Belt), impact having taken place when one asteroid collided with another. (It might be noted that fragments in asteroidal breccias become cemented together from the shock of later impacts, which then "stick" the previously generated fragments together).

A large bass (*Prisicara*) jumping from the impact produced lake (*Cratersee*) of the Steinheim structure after the impact crater filled with water and became a fish-filled lake. *Artwork by William Brownfield.*

Impression of a reed (or large grass stem) that grew along the edge of the crater-formed lake of the Steinheim structure: Grass and reeds (cattails) are relatively recent in their appearance—first appearing in the Miocene Epoch. (Value range F).

These rocks were formed from sediments "scraped" from the Pacific Ocean floor and then "plastered" to the western edge of North America. This scraping process is a consequence of the eastward movement of a small tectonic plate as well as the westward movement of North America, both movements being part of plate tectonics. The pipe (or trumpet) fish shown here also have an association with West Coast tectonic processes that "plastered" diatom-rich sediments containing the pipe fish against the west side of California. Neogene sediments involved with tectonic activity like this generally become hard and consolidated in contrast to those that occur in areas of little tectonic activity, like Florida, where they are soft and poorly lithified.

Centriscus strigatus, Gunther, 1861: A group of fossil razor or trumpet fish. These interesting fish come from a zone known for its well-preserved fossils, which includes these fish. Marecchia River Formation, Miocene, Poggio Berni, Rimini Province, Italy. (Value range D).

Gyraulus trichiformis: Fresh water gastropods (snails) preserved in fresh water limestone deposited in the Steinheim "impact-formed" lake. These small fossil snails occur in vast numbers (Schneckensanden) associated with sediments of the Steinheim impact structure. (Value range G, for group).

Fish, Especially Bony Fish (Teleosts)

Fish can be quite desirable fossils (especially if articulate and complete) in Neogene rocks, as well as in older strata. Neogene fish generally are similar, as are most other Neogene fossils, to fish living today. Deep-sea fish like *Lampauyctus* occur in siliceous sediments, which make up parts of the coast ranges of California.

Sygnathus (Hipposygnathus) acus, Linne, 1758: Same specimen as on right—different lighting.

Centriscus strigatus, Gunther, 1861: Compressions of two specimens of razor or paddlefish on a bedding plane. Excellent fossil fish have been known since the early eighteenth century to occur in these grey clays, which crop out along the Marecchia River of Italy. Marecchia River Formation, Miocene, Poggio Berni, Rimini Province, Italy.

Sygnathus (Hipposygnathus) acus, Linne, 1758: Long nose pipefish. A compression of this peculiar fish is well preserved in Miocene lagoon sediments (grey clay). This fossil fish originally was described by Carlos Linnaeus (Carl Linne) in 1758. Linnaeus was the originator of the Linnaean system used in classifying the vast and complex categories of organisms that constitute biology. The occurrence of these fossils has been known for well over 200 years. (Value range D).

Sygnathus acus, Linne, 1758: Long nose pipefish. Two complete compressions of this relative of the seahorse. Marecchia River Formation, Miocene. Poggio Berni, Rimini Province, Italy. (Value range D).

Lampauyctus sp. (lanternfish): Deep-sea fish preserved as compressions in diatomite. Puente Formation, Los Angles Co., California. (Value range G, single specimen).

Seahorses (*Hippocampus* sp.): Fossil seahorses come from the same Lower Pliocene outcrops near Poggio Berni as do the pipefish, but are considerably rarer than are the related fossil fish.

Deep Sea Fish

Lampanyctus petrolifer (lanternfish): This is a deep-sea fish preserved in diatomite. Deep-water sediments, "plastered" onto the western edge of North America by sea floor spreading occur in the state of California and in Baja, California. These sediments sometimes contain impressions of deep-sea fish, some of which had glowing extensions as a consequence of their living in deep waters below the photic zone. Miocene, Puente Formation, Los Angeles Co., California. (Value range E).

Mallotus villosus (Atlantic capelin): A calcareous concretion containing the impression of this fish from Pleistocene sediments deposited in the Champlain Sea. Zones that yield these concretions occur in the bed of the Ottawa River near Ottawa, Ontario, Canada. These fish bearing concretions came from sediments of the Champlain Sea, an extension of the Atlantic Ocean, which covered parts of eastern North America toward the end of the ice age. Similar fish bearing concretions, most of which are hard and stone-like, are found in geologically young sediments at other places in the northern U.S., Canada, and in Greenland. Such geologically young fossils have been cited by young earth creationists to "**prove**" that the earth is only 6,000 to 10,000 years old! (Value range E).

Mallotus villosus: Same specimen, different lighting conditions.

> northern drift of Ireland and Scotland. Some of the concretions of fine clay, more or less calcareous, met with in New Hampshire, in this "drift" on the Saco River, thirty miles to the north of Portsmouth, contain the entire skeletons of a fossil fish of the same species as one now living in the Northern Seas, called the capelan (*Mallotus villosus*), about the size of a sprat, and sold abundantly in the London market, salted and dried like herrings. I obtained some of these fossils, which, like the associated shells, show that a colder climate than that now prevailing in this region was established in what is termed "the glacial period." Mr. Hayes took me to Kittery, and other localities, where these marine organic remains abound in the superficial deposits. Some of the shells are met with in the town of Portsmouth itself, in digging the foundation of houses on the south bank of the river Piscataqua. This was the most southern spot

Mention and explanation of the fossil fish *Mallotus villosus* found in concretions occurring in glacial lake sediment. These fish bearing concretions occur at a number of locations, especially in those areas where sediments deposited in the Champlain Sea occur. The Champlain Sea was an extension of the Atlantic that covered parts of eastern North America near the end of the ice age. From Charles Lyell's *A Second Visit to the United States of North America*, 1846-'47, John Murray, publisher.

Deep-sea fish: These fossil fish impressions occur in siliceous, hard shale, which forms sea cliffs in parts of southern California. Monterey Formation, Miocene, near Big Sur, California. (Value range F for group).

Pipe or trumpet fish: A group of Pliocene pipe fish from (possible) deep-sea sediments near Buellton, California. Pipefish appear to occur somewhat widely in Miocene and Pliocene sediments around the globe. (Value range F for single specimen).

Deep-sea fish? Monterey Formation, Miocene, Big Sur, California. (Value range G).

Pipefish: A single specimen of this pipefish from deep sea sediments near Buellton California. Specimens from this locality often are not as well preserved as similar ones from Italy. (Value range F).

Chapter Six
Neogene Amphibians and Reptiles

Amphibians

Amphibians of the Neogene generally are similar to those of today. Frogs, toads, and salamanders form the main groups of Late Cenozoic amphibians. These are generally rare fossils as the bodies of these lovely animals are fragile and are preserved under only the most ideal conditions.

Rana basaltica: Same frog impression, but with different lighting. Frogs are very desirable fossils—savor this one! Miocene lake sediments, Shandong Province, China.

Rana basaltica: An impression of a very fine frog. Fossil amphibians generally are rare—most amphibians being delicate and leaving no fossil record (or at best a poor fossil record, particularly in the Cenozoic Era). Fossil frogs, especially articulated ones like this are rare (and desirable) fossils. Amphibian fossils are almost totally absent in most sedimentary environments, which preserve mollusks or arthropods. This is the impression of a frog preserved in Pliocene lake bottom sediments. Lake bottom sediments offer the best opportunity to preserve and yield fossil amphibians like this mudstone impression. Miocene, Shandong Province, China. (Value range E).

Close-up of the fossil frog.

Rana sp. Modern frog: Unless buried in very fine sediment, very quickly, such a lovely, little, but fragile body as seen with this fine frog would have little likelihood of becoming a fossil.

Reptiles

Turtles: Lake and river sediments, relatively common in the Cenozoic, can contain the fossil shells of turtles. Turtles being the most abundant reptiles found as Cenozoic fossils. Despite the abundance, complete fossil turtle shells are quite desirable—shell fragments … less so!

Tortoises: Large land turtles (tortoises) in the Miocene and Pliocene epochs lived at relatively high latitudes. They also lived (at times) during the interglacial stages of the Pleistocene. Populations of tortoises migrated to higher latitudes as the climate warmed after a period of glaciation in much the same way as warmer climate animals (like armadillos) are migrating today toward higher latitudes as a consequence of global warming. Today only relic populations of these large tortoises exist, usually on islands like the Galapagos of the eastern Pacific. During the Neogene, large tortoises lived over a large part of the earth's land areas, sometimes extending into latitudes that today would be way too cold for their survival. (Large tortoises cannot burrow into the ground and hibernate as smaller ones can).

Testudo sp. Giant Miocene tortoise: Giant tortoises lived worldwide during the Miocene and Pliocene Epochs—fossils of them (or at least fragments) can locally be common fossils in some parts of the world, including the southern states of the U.S. Complete specimens, or nearly complete ones like this one, preserved in tuffaceous sediment (volcanic ash beds) from the Miocene of Nebraska, are not common. This tortoise was buried by volcanic ash associated with volcanism of the Miocene Epoch, which came from a series of massive volcanic eruptions originating in the region of what is now Yellowstone National Park. (Value range C).

Large tortoises like this live today on a few islands, like the Galapagos Islands, where they represent a relic population of what were, during most of the Neogene, widespread animals—at least at lower latitudes. Except for isolated island locations, these large reptiles went extinct during or at the end of the Pleistocene Epoch.

Testudo sp. A fragment of a large Galapagos-like tortoise: Shell fragments of these large reptiles locally can be common fossils in Florida, Texas, and other southern states. The age of such fragments is generally Pliocene or Pleistocene (from the interglacial periods of the ice age when it was warmer than it is today). (Value range G, single fragment).

Portions of large tortoises: A recognizable portion of one of these large tortoises is a more desirable fossil than are non-descript fragments. Pliocene, Santa Fe River, Florida, except for the white fragment being held, which is from the Fort Worth area of Texas. (Value range F).

Thick (two inch) fragment of an especially large tortoise: Pliocene? Continental shelf sediments, Florida. Numerous fossils like this are found by diving and accessing Neogene sediments, which crop out below sea level along the Florida coast. Notice the (modern) barnacles (white stain) attached to the tortoise shell fragment. (Value range F).

More fragments of large tortoise shells. These interesting fossils are readily available as large numbers of them come from Florida rivers, where they are often collected by scuba diving.

More fragments of large tortoise shells. Pliocene, Florida.

Geochelone sp.: Large tortoises ranged over southern North America during the Pliocene and in some areas during the interglacial stages of the Pleistocene. Most of the fragments shown here came from the Dallas-Fort Worth area of Texas—coming from deposits along the Trinity River. (Value range F).

Titanocheyles sp.?: More dermal scutes from large tortoises. Dermal scutes of various kinds from a variety of animals occur as fossils. These include sharks, turtles, and even mammals (sloths), and can be more frequently found as fossils than are other parts of these organisms. (Value range F for group).

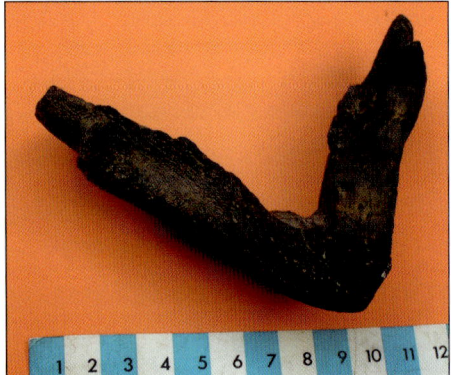

Turtle beak—Pliocene, Florida. (Value range E).

These large flat scutes come from the legs of giant turtles. Pleistocene, Santa Fe River, Florida. (Value range F for group).

Dermal scutes from large Pliocene or Pleistocene tortoises: These bony scutes originate from the appendages of large tortoises where they are set in the animal's skin. Santa Fe River, Florida. (Value range F, for group).

Tortoise in loess: Miocene, China. Loess is a type of wind blown sediment that is usually associated with the ice age. Miocene loess occurs in China that sometimes contains well-preserved fossils. These are from animals that were overwhelmed by the fierce dust storms that blew and deposited this fine rock flour. (Value range D).

Crocodilians

Fossil remains of crocodilians are found worldwide in Miocene and Pliocene strata. Cooling conditions of the Pleistocene Epoch reduced their range considerably so that today they live only in limited tropical or subtropical areas.

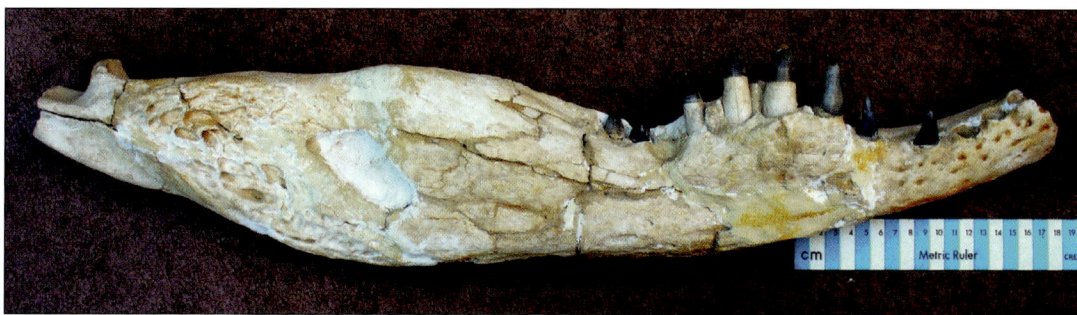

Gavialosuchus americanus: Crocodilian jaw. Crocodiles lived worldwide during the Early Neogene, but like the large tortoises of the same time, their range was considerably reduced during the ice age and by the end of the Pleistocene their populations became very localized. They were no longer the cosmopolitan animals that they had been before the ice age. This crocodile jaw came from Miocene Bone Valley phosphate deposits, Polk Co., Florida. (Value range D).

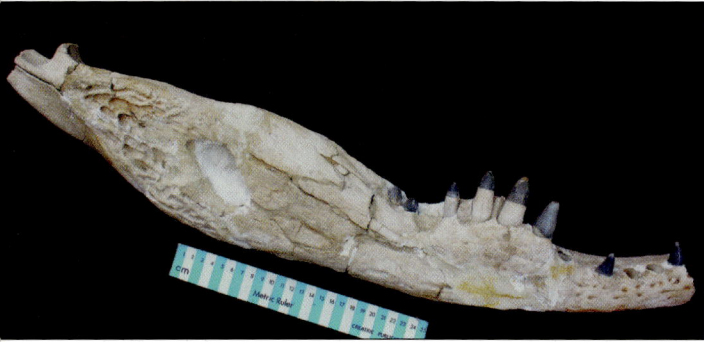

Side view of the crocodile jaw shown in the previous photo.

Gavialosuchus americanus: Teeth of crocs can be relatively common fossils in Neogene strata. These came from the Miocene Bone Valley Formation of Polk County, Florida. (Value range H).

Portion of a crocodile jaw: Pliocene, Santa Fe River, Florida. Croc jaw sections like this one are desirable fossils. (Value range E).

Alligator mississippiensis: Alligator scutes. Alligator scutes are more regularly shaped than are those of crocodiles. Compare this with the below scutes from crocodiles. Pleistocene, Polk County, Florida. (Value range G for group).

Coprolites

Coprolites believed to be of reptilian origin have become widely distributed among collectors of fossils. Replaced by iron oxide (limonite), these "turd-like" trace fossils are generally found in Miocene strata of Oregon and Washington. Some paleontologists dispute their reptilian origin and some may be of mammalian origin. Both why they are found in such concentrations and what the conditions surrounding their replacement with iron minerals (petrifaction) remain a mystery.

Coprolites: These mineralized fossil excreta (turds) are supposedly of reptilian origin. They are replaced (mineralized) with hydrous iron oxide (limonite). Coprolites like these from Oligocene and Miocene strata of Oregon and Washington have been widely distributed and are quite popular with rockhounds—they are also popular with children who are amused by the "potty humor." They have often been stated as being from dinosaurs—however, this is impossible! Dinosaurs went extinct 30 million years before these were "pooped out" by large (presumed) lizards. These coprolites have also been suggested to have come from manatees. (Value range F, single specimen).

Gavialosuchus americanus: Crocodilian scutes. Crocodiles lived throughout the world during the Neogene. These scutes are from the Miocene of Alachua County, Florida. (Value range G for group.)

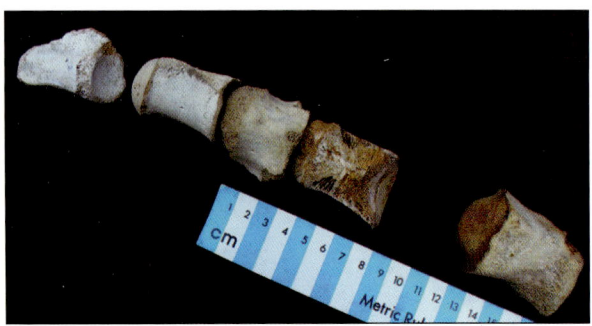

Gavialosuchus sp.: A set of crocodile vertebrae, Miocene, Alachua County, Florida. (Value range F for all).

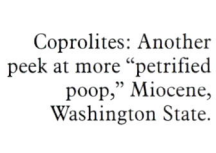

Coprolites: Another peek at more "petrified poop," Miocene, Washington State.

Coprolites from Madagascar: These iron-oxide-replaced? coprolite-like objects appear similar to those from the Pacific northwest of the U.S.; however, these have recently (2006) entered the fossil market from Madagascar. They appear similar to those from Oregon and Washington, however many exhibit a strange granular texture—a texture suggesting that they at one time may have been composed of pyrite and that the large granules composing them are pseudomorphs after that mineral. (Value range G).

More coprolites from Madagascar.

Bibliography

Amstutz, G. C., 1958. "Coprolites: A review of the literature and a study of specimens from southern Washington." *Journal of Sedimentary Petrology*, Vol. 28, No. 4, pp. 498-508.

Ivanov, Martin, Stanislava Hrdlickova, Ruzena Gregorova, 2001. *The Complete Encyclopedia of Fossils*. Rebo Publishers, The Netherlands. ISBN 90-366-15003.

Parker, Steve, 2007. *The World Encyclopedia of Fossils and Fossil Collecting*. Lorenz Books, Anness Publishing Ltd., Hermes House, London.

Walker, Cyril and David Ward, 1992. *Fossils, Eyewitness Handbooks*. D-K Publishing. ISBN 1-56458-071-7

Chapter Seven
Miocene and Pliocene Mammals

The Miocene and Pliocene epochs are notable in their worldwide diversity of mammalian fossils. Represented here is but a small sampling of this diversity.

Miocene Mammals of the Molasse

A representative Miocene mammalian fauna occurs in clastic sediments derived from the uplift of the Alps and deposited to the north between Switzerland and Bavaria—a sequence of sediments known as the molasse. Its fauna is similar to Miocene mammalian faunas found elsewhere over the globe, including those found in Nebraska and eastern Wyoming, as well as in the John Day fossil beds of Oregon.

Plesioceratherium sp.: This jaw section is from the molasse, a thick sequence of sediments eroded from uplifting of the Alps, a process which took place during the late Cenozoic. Miocene, Gunzburg, Bavaria, Germany.

Beaver incisor: Many Miocene animals are similar to those of today. This is the incisor of a normal-sized beaver. Throughout the Late Cenozoic, there is a pattern of mammals becoming progressively larger until the large mammals of the Pleistocene are reached (Pleistocene Megafauna). Compare this incisor to that of the large beavers of chapter ten. Molasse deposits, Gunzburg, Bavaria. (Value range G).

Plesioceratherium sp.: Jaw sections of an early rhinoceros. Miocene, Molasse, Gunzburg, Bavaria, Germany. (Value range E for group).

Teeth of Gomphotheres

Teeth and bones of probosidians (elephant-like mammals) can be locally abundant, conspicuous, and desirable fossils. The stegomastodon was an ancestor to the Pleistocene mastodon and mammoth. A goodly number of these probosidian's oddly shaped teeth came onto the fossil market in the late 1990s. These teeth originated from gravel pits in Indonesia and have become widely distributed among museums, dealers, and collectors. The stegomastodon is a type of gomphothere—gomphotheres are an earlier form of the mastodon.

Large, single Stegomastodon tooth: Fossil teeth of the Stegomastodon come from Pliocene stream deposits and are collected as cobbles which occur in the bed of the Solo River as well as in other rivers of Indonesia. Large numbers of this ancestor to the mastodon must have lived in parts of Indonesia, as their water worn teeth can be fairly common fossils in this part of the island of Java. Stegomastodon teeth like these are also found in South America. (Value range D).

Highly worn tooth of Stegomastodon: Well worn teeth like this one came from individuals who lived to a ripe old age—their teeth having been ground down by years of eating gritty vegetation. (Value range D).

A different view of the stegomastodon tooth shown in the previous photo.

Tooth of juvenile stegomastodon: Pliocene, Java, Indonesia. (Value range E).

Stegomastodon sp. teeth: This is a group of these interesting fossil teeth which occur in gravel beds composing terrace deposits of the Solo River, a major stream on the island of Java, Indonesia. Excavations into these gravels for road building materials have produced considerable numbers of these probosidian teeth. Some vertebrate paleontologists have been distressed by these and similar fossils ending up at the Tucson show and falling into the hands of collectors. They allege that collectors, in their willingness to purchase such fossils, drive up their prices and prevent such material from going to science and academic investigation. What is **not being considered** is the condition under which such fossils go onto the fossil market. Many fossil as well as mineral specimens come from mining operations! The fact that someone would give a monetary incentive (viz. money) for these fossil teeth is what usually serves to salvage them in the first place—otherwise they would have gone into piles of large sized gravel. Collectors and dealers willing to "fork out" hard cold cash are what, in many cases, prevent educationally and scientifically valuable fossils and minerals from ending up either being destroyed or reburied. (Value range D for group).

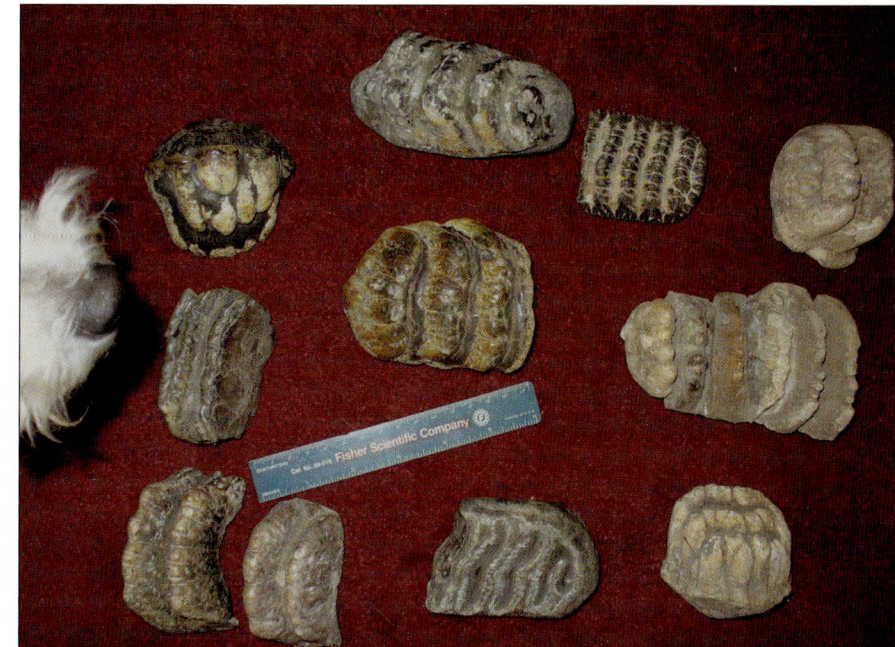

Rhinoceros

Rhinos are generally thought to be an animal endemic to Africa, which is the case today; however, the rhinoceros in the Miocene and Pliocene Epochs lived over a large portion of the globe. The teeth of these animals are found in terrestrial stream deposits in many parts of the world, including North America.

Stegomastodon teeth with distinct, unworn cusps: Pliocene, Java, Indonesia.

Rhinoceros teeth, Miocene, China: The rhinoceros occurred worldwide during the Miocene Epoch. (Value range E).

A cast of a reconstructed stegomastodon displayed at the 2009 MAPS Expo fossil fair.

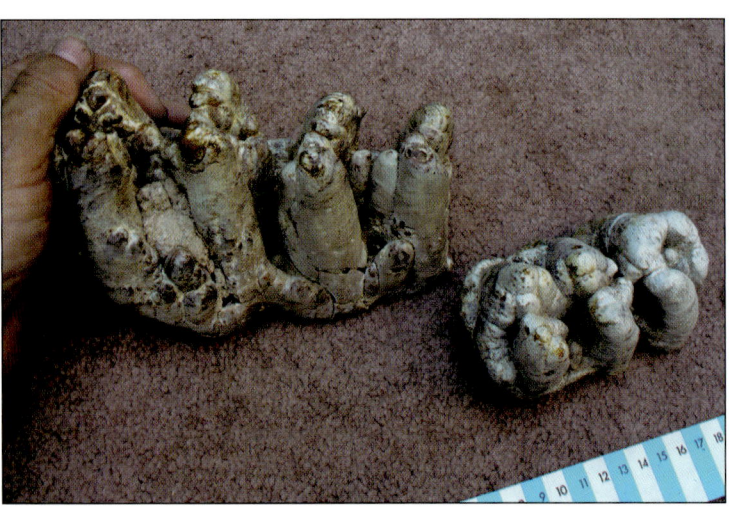

Platybelodon sp. Gomphothere teeth: These teeth came from a shovel-tusk elephant—a gomphothere. This is a worldwide occurring Pliocene proboscidian. Late Pliocene, Ganshu Province, Northwest China. (Value range E).

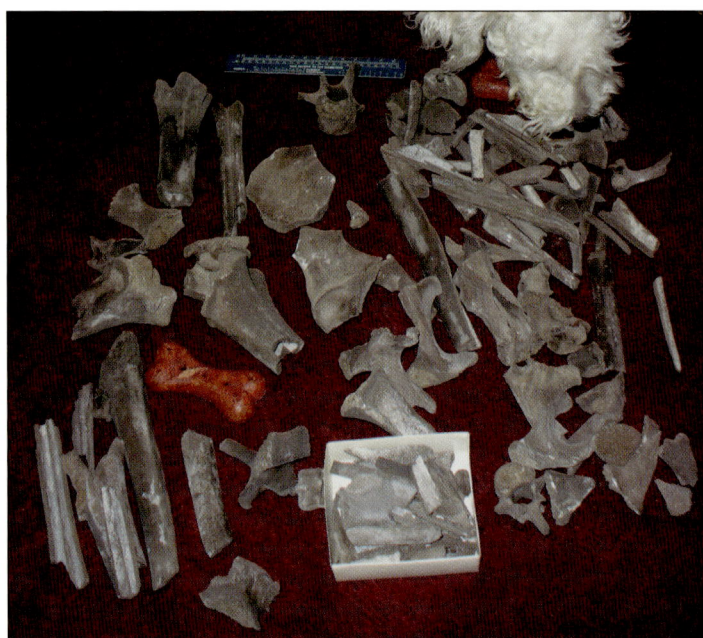

Horse: The horse evolved in North America where it is represented by a variety of genera and species. These bones found in an excavation for a shopping center are those of a small Pliocene horse. The skull was not found, unfortunately—however reconstruction of the animal from these bones would present a paleontological exercise—one that requires a good knowledge of anatomy and one which would give a pre-med or anatomy student some excellent "hands on" experience. This is what most vertebrate fossils appear like before preparation—a pile of fossil bones! (Value range F, unprepared).

Carnivores

Carnivores are relatively uncommon fossils, particularly those of the cat and dog families, especially when their abundance is compared to that of herbivores. Herbivores, being relatively stupid and having larger populations, become trapped in situations that kill them and produce fossils. Carnivores, on the other hand, usually avoid such situations; however, there are exceptions, like the La Brea tar pits of southern California. Also compared to herbivores, their populations were (and are) relatively small so that overall fossils of Neogene carnivores are relatively rare (and desirable).

Basilosaurus (Zeuglodon) cetoides Owen: This is a cast of the jaw of a primitive whale found in late Paleogene (Upper Eocene) strata of Louisiana, Mississippi, and Alabama. Remains of whales and other cetaceans are somewhat common fossils in marine Cenozoic strata of the Gulf Coast Series and this is one of the earliest of such occurrences. The genus *Basilosaurus* (saur=lizard) indicates that this fossil, when first described in the 1840s, was considered to be an extinct reptile. It was first described by Sir Richard Owen of England—an early vertebrate paleontologist who originated the concept (and name) of dinosaurs.

Dire wolf skull: The skull of an early canid preserved in Miocene loess, China. (Value range E).

Marine Mammals

Fossils of marine mammals are some of the most commonly found mammal remains in Miocene and Pliocene strata. Their bones can be large and some of them have attracted considerable interest as well as controversy, such as the case with those of the Hydrarchos (the water king) of the nineteenth century. Displays of some of these weird reconstructions rivaled the show(s) of P. T. Barnum and had a full compliment of followers, including "snake oil" salesmen and sellers of Mrs. Winslow's Soothing Syrup. Various strange looking bones, including the ear bones of whales, are also frequently found in Miocene and Pliocene marine strata. Also in a category by itself are the bones and teeth of sea cows (manatees).

Whale vertebrae: Miocene. Whale vertebrae are found in Neogene sediments (and in early Cenozoic rocks as well) in many parts of the southern United States. One of the most notable was the occurrence of the vertebrae of what were originally described as *Zeuglodon*—*Zeuglodon* vertebrae were found to be the vertebrae of an extinct whale, the bones of which were found in Alabama and the subject of a controversy discussed in Charles Lyell's *Second Visit to North America* 1846-'47. Lyell discusses large vertebrae (larger than those shown here) found in abundance near Clarksville, Alabama, in the early 1840s. The giant whale vertebrae of *Zeuglodon*, which caused such a sensation in the mid-nineteenth century were reconstructed as a huge monstrosity and placed on display as the **Hydrarchos** in a competitive exhibit to those of P. T. Barnum. Today these Paleogene whale vertebrae are infrequently found, in contrast to their relative abundance in Neogene strata. Information on fossil whales and Zeuglodon are placed here in the Neogene book, rather than in the Paleogene text that precedes this work, for that reason.

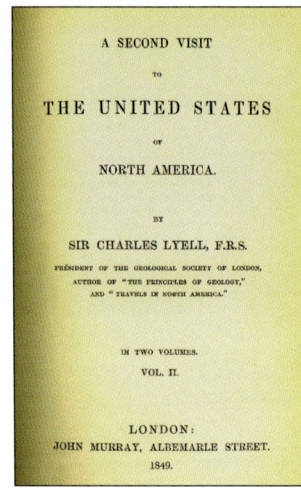

Title page of Charles Lyell's *Second Visit to the United States of North America*, 1846-'47.

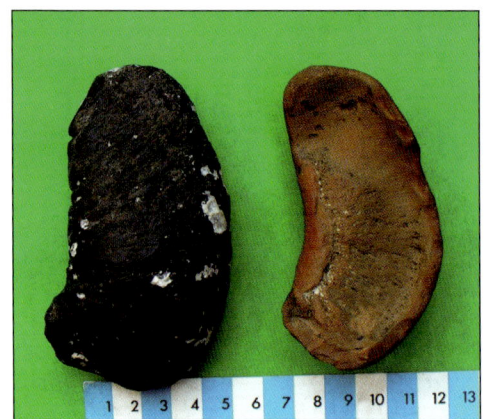

Whale tympanic (ear) bones: These peculiarly shaped bones are utilized by whales to assist in their acute, underwater hearing. For some reason, these ear bones are particularly common fossils in Miocene strata of the eastern U.S. These specimens came from the Miocene of Maryland. (Value range G, single ear bone).

Discussion on fossil whales by Charles Lyell in his *Second Visit to North America*, 1846-'47. Here he discusses *Zeuglodon* or *Basilosaurus* as well as the *Hydrarchos* of Albrecht Koch who also reconstructed another monstrosity known as the Missourium from mastodon bones found south of St. Louis, Missouri (see chapter ten). The large fossil whale vertebrae that formed the basis of Hydrarchos (The Water King) as well as other mid-nineteenth century paleontological puzzles are discussed in this work by Charles Lyell. Lyell is often considered to be the founder of modern geology (or at least he put it on a firm basis), with its emphasis on fossils and strata, a necessary prerequisite to Charles Darwin's natural-selection-based, evolutionary theory.

Whale tympanic ear bone, Miocene Maryland.

Desmostylus hesperius: Teeth of a sea cow. Sea cows, also known as manatees, provide some of the most frequently found vertebrate fossils in Cenozoic marine sediments. These teeth are found in marls of the Miocene, Santa Margarita Formation near Fresno, California, a rock unit that has yielded a number of these distinctively shaped teeth, which resemble a six-pack of beer. Sea cow teeth are not common fossils, but their vertebrae, by contrast, can be. (Value range E).

Here is mentioned the monstrosity which Albrecht Koch called the Hydrarchos or "Water King," which he exhibited in New York in 1845. *Zeuglodon* is now the state fossil of Alabama, but few remains are found today compared to the numbers that surface from Neogene strata. The Dr. Emmons mentioned is Ebenezer Emmons who was one of the pioneer geologists of North America—among his various accomplishments being the first geologic map of an entire state (New York).
Here Lyell states proof that the large vertebrae and associated fossil teeth were really from cetaceans (whales).

Claiborne, Alabama, is located above a steep bluff on the Alabama River. In the nineteenth century, it was a steamboat landing with a long set of stairs leading to the top of the bluff—a facility that helped Lyell in collecting fossil mollusks found here in Eocene clay beds of the Claiborne Group. The Tombekbee River is now known as the Tombigbee River, some of the fossils of Chapter Four of this work are from its bluffs.

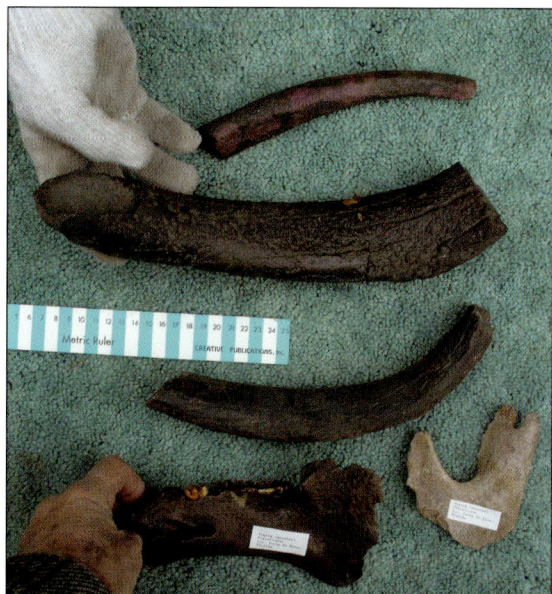

Manatee ribs: Miocene, Florida. Mineralized ribs of manatees (sea cows) are common fossils in Miocene marine strata of Florida. Phosphate rock, mined in central Florida (Bone Valley Formation) can produce them in abundance. Manatee ribs also occur in river deposits. Bone Valley Formation, Lakeland, Florida. (Value range G, single rib).

Camel tooth: A highly mineralized and worn camel tooth found on the Grand River, northwest Missouri. The presence of Pliocene (Pre-Pleistocene) sediments in northwest Missouri is likely, as similar age strata have been found in adjacent Iowa. Soft sandstones that crop out along the Grand might be of Pliocene or Miocene age, however fossils have not been found in them. These water-worn fossil teeth are found on the gravel bars of the river and may have come from Pre-Pleistocene sediments or they may have been carried down from Iowa. (Value range G).

Porpoise vertebrae: Beside manatee remains, the vertebrae of porpoises can be common fossils in Miocene strata. Pungo River Formation, Lee Creek Mine, Beaufort Co., North Carolina. (Value range G).

Scelidotherium capellini: Sloth claw. A distinctive claw type from a South American ground sloth. These giant ground sloths of South America were first reported on and brought to the attention of science in Charles Darwin's accounts of his voyage of the *Beagle* in 1838. Pliocene, Tarija, Bolivia.

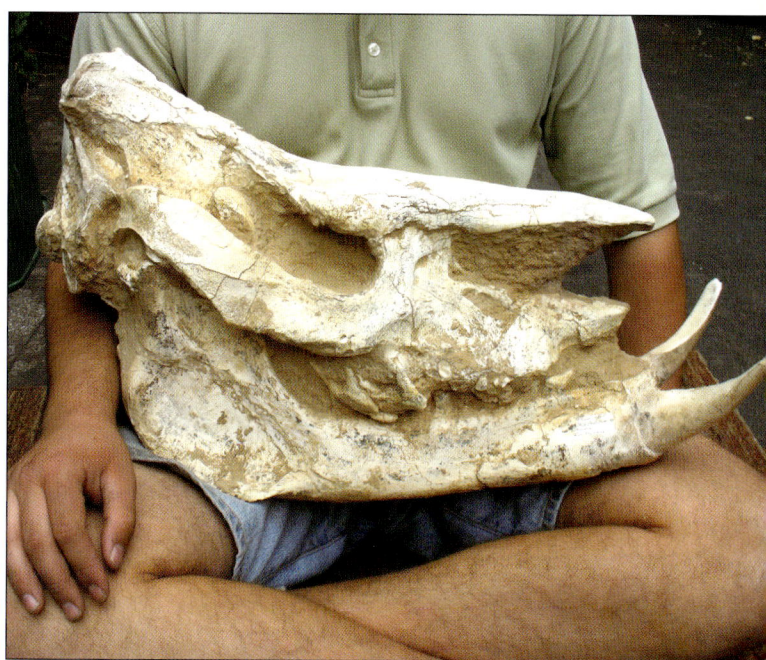

Chilotherium sp.: A large **rhinoceros-like** skull found and preserved in Miocene loess. Similar rhinoceros-like herbivores lived in North America and occurred worldwide during the Miocene. Their occurrence in North America is associated with volcanic ash beds of the central high plains where the animals appear to have been overcome from falls of volcanic ash. The animal that bore this skull similarly may have been overcome by the wind blown silt associated with Miocene dust storms. Miocene, Linxia Basin, Gansu Province, China. (Value range A).

QUATERNARY PERIOD

Chapter Eight
Invertebrates of the Pleistocene Epoch

Pleistocene Corals and Mollusks

The Pleistocene Epoch (known also as the "Great ice age") was a time in the earth's recent geologic history when large parts of the planet's land masses were covered by continental glaciers, especially in the Northern Hemisphere. The Pleistocene consists of four major periods of glacial buildup sandwiched between periods when the earth's climate was warmer at times than it is today. These warm periods are known as the interglacial stages. Cooler temperatures associated with ice build-up on the continents resulted in entire ecosystems shifting toward lower latitudes. When the ice melted and warmer conditions again prevailed, the warmer climate ecosystems migrated to higher latitudes. Thus, in any particular region, but especially in the Northern Hemisphere, Pleistocene strata record a sometimes-confusing mix of cold climate life forms alternating with those of a warmer climate. It was during the Pleistocene that humans migrated over the globe from our place of origin in Africa—the African continent having the best (and the oldest) fossil record of hominids. The appearance of humanity in Africa is "possibly" one of the consequences of the ice age. Pleistocene continental glaciation, although not directly affecting Africa, drastically changed its climate from a wet-tropical one to one of a dryer, more savanna-like climate. This climate change may have been responsible for a monumental event, the evolution of human intelligence. **The Pleistocene**, with its ice ages and continental glaciation, appears in **many ways to have been a unique part** of geologic time!

Pleistocene Marine Fossils

During periods of continental glaciation, vast quantities of water normally found in the world's oceans were tied up forming the huge masses of ice that composed the continental glaciers. These glaciers covered major parts of the earth's landmasses, especially in the Northern Hemisphere. This ice, which ultimately came from the oceans, dropped sea levels over 1,500 feet when the ice was at a maximum. During the interglacial stages, this glacial ice melted and sea level rose—inundating the continental shelf and in warmer climates forming limestone at the margins of the continents. These Pleistocene limestones, having formed in warm water, can be full of marine fossils—fossils like mollusks and echinoderms, which often are well preserved and resemble the bleached shells found along modern sea shores.

Outcrop of Pleistocene limestone near Montego Bay, Jamaica: White limestone containing fossil corals and large conch shells crops out along the coastal parts of Jamaica as well as on other islands of the Caribbean.

Corals occur frequently in Pleistocene limestone of southern Florida (especially in the Florida Keys), as well as in many coastal regions of the Caribbean. Corals thrived during the interglacial stages, where Pleistocene limestones of the Caribbean and Central America can be full of fossil corals like these.

Pleistocene limestone containing crystallized fossil coral heads, western Jamaica. These corals, known as hexacorals, generally are the same types as those living today.

Large fossil coral head in Pleistocene limestone near Montego Bay, western Jamaica: Corals like this grew in shallow waters at the margins of most Caribbean islands as well as southern Florida during the interglacial stages of the Pleistocene when sea level was as high or higher than it is today.

Pleistocene limestone can contain fossil coral heads that are a part of the limestone deposited by the higher sea levels that existed during interglacial stages of the ice age.

Brain coral from western Jamaica: These occur in coastal limestone outcrops where they can locally be abundant on many islands of the Caribbean. (Value range G).

These are two nice and large corals from the Pleistocene Caloosalatchee Formation, a soft fossiliferous marl or limestone that occurs in central Florida. These were found along the Caloosalatchee River. (Value range G, single specimen).

Manicina areolata: Caloosahatchee Formation, Florida. (Value range H).

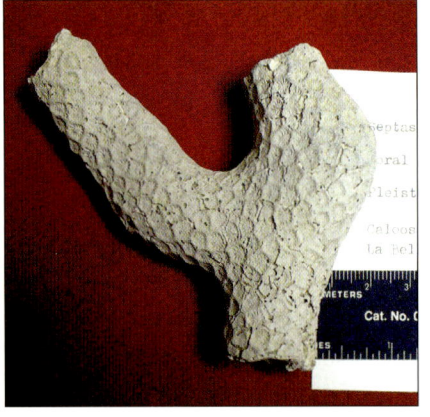

Septastrea crassa: Caloosalatchee Formation, Caloosalatchee River, central Florida. (Value range H).

Manicina areolata: Caloosahatchee Formation, central Florida. (Value range H).

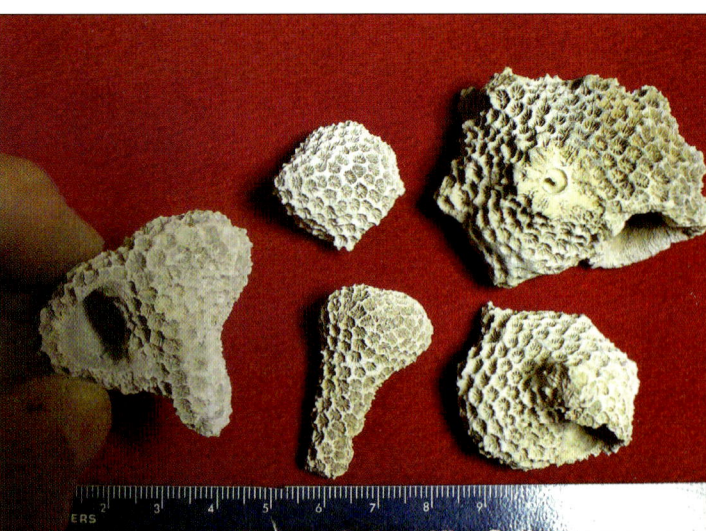

Septastrea marylandica: This coral is unique to the Caloosahatchee Formation of the Caloosahatchee River of Florida. It grows **over** the shells of marine gastropods, entirely covering the outside of abandoned shells. (Value range G for group).

Acylonaria sp.: A commonly found Pleistocene coral, central Florida. (Value range H).

Marine Gastropods (Snails)

Pleistocene marine gastropod faunas are large and quite similar to those of today. Most of the species found as fossils are still living today, although those which lived during the actual periods of glaciation represent colder water species now found living at higher latitudes than those of Florida or the Caribbean. The sea now covers much of the limestone deposited at low latitudes during periods of glaciation. Sea levels rose with the melting of the continental glaciers.

Representative Pleistocene coral specimens of a reasonably collectable size are shown here. Many fossil corals of the Caribbean and southern Florida can be quite large. These large masses of fossil coral are sometimes used in various ways as ornamental rock with the pattern of the coral often emphasized.

Large unidentified marine gastropods like this occur in the Caloosahatchee Formation of Florida. These large guys are rare; most marine mollusks found in the marls are smaller. (Value range F).

Lightning whelk, Nasha Formation, Okeechobee, Florida. (Value range H, single specimen).

John Stade with a large specimen of the marine gastropod *Busycon contrarium*, a left handed "lightning whelk". These sinistral coiling (counterclockwise) gastropods have been given their own "new" genus and species (*Sinistrofulgur sinistrum*). Caloosahatchee Formation, Caloosahatchee River, Florida. (Value range F).

Here is a group of Caloosahatchee Formation fossil mollusks in the specimen drawer of a well-organized collection. *Courtesy of John Stade*.

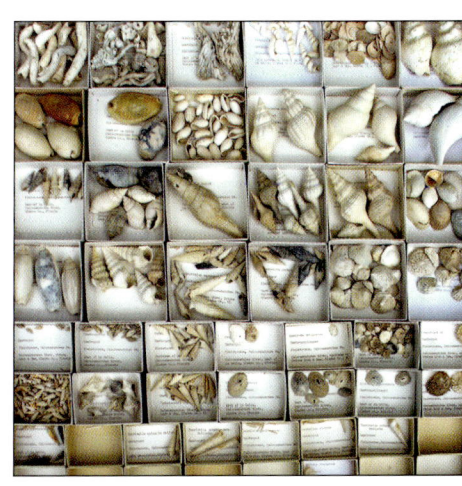

Turbinella regina: Caloosahatchee River, LaBelle, Glades Co., Florida. (Value range G).

Pleistocene Fossils of Florida—Especially of the Caloosahatchee Formation

Pleistocene limestone covers large parts of Florida—especially making up parts of the underlying rock of peninsular Florida, particularly in the southern part of the state. Here marl (soft, chalky limestone) can be full of fossil shells; sometimes these shells being beautifully preserved. The Caloosahatchee Formation is particularly well known for its marine shells and contains one of the most extensive molluscan faunas of this age in the world—a fauna deposited in water, which was relatively warm. The Caloosahatchee fossils illustrated here were collected and identified by John H. Stade—a collector in the St. Louis, Missouri, area who has spent a considerable amount of time collecting Cenozoic fossils in the southern states, particularly in Florida.

Pleistocene marine gastropod faunas are large and are similar to those of today. Most of the **species** found as fossils are still living today, although those that lived during periods of glaciation represent cold-water species, which are now found at higher latitudes. Many of the limestone beds deposited during periods of glaciation are now covered by hundreds of feet of water. Sea levels have risen with the melting of continental glaciers to the levels of today, which were also the levels experienced during the interglacial periods.

Crucibulum multilinentum: Caloosahatchee River, LaBelle, Glades Co., Florida. (Value range F for group).

Siphocyprae carolinensis: Caloosahatchee River, LaBelle, Glades Co., Florida. Value rane H for group).

These peculiar gastropods resemble calcareous worm tubes. The generic name *Vermicularia* (verm = worm) refers to this characteristic. Shells of the gastropod *Vermicularia* also cluster together like a mass of worms.

Essay by John Stade—St. Louis, Missouri

Most tourists who go to Florida in the winter spend at least a day or two on the beach collecting seashells. We go to Florida to collect shells also, but we head inland, not toward the beach. We are looking for fossil shells—Pliocene and Pleistocene in age!

Vermicularia recta: A peculiar worm-like gastropod shell that congregates to form large masses resembling worm tubes. Caloosahatchee River, LaBelle, Glades Co., Florida. (Value range G).

Vermicularia recta with end view of shell mass. (Value range F).

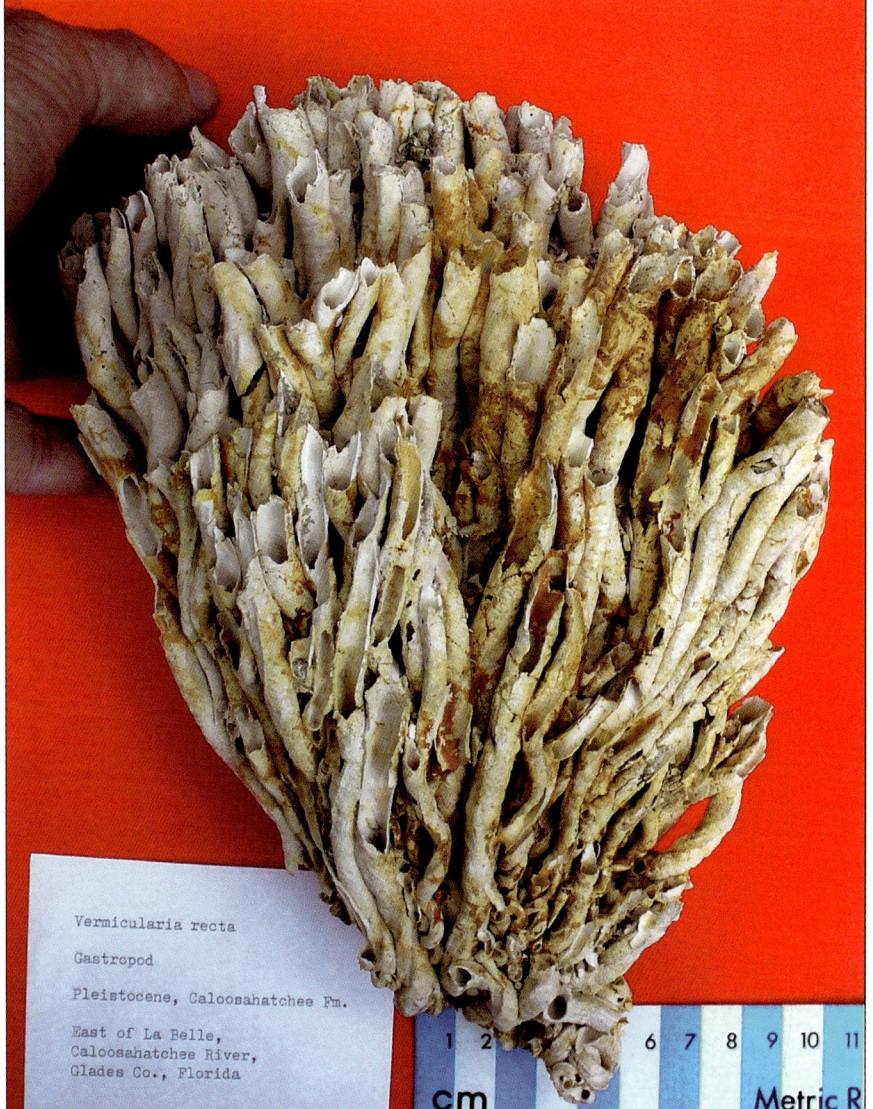

Vermicularia recta: Side view of mass of *Vermicularia* shells. LaBelle, Caloosahatchee River, Glades Co., Florida. (Value range G).

One of our favorite Florida collecting areas is along the Caloosahatchee River. What used to be the Caloosahatchee River has been dredged into a fairly straight, relatively deep canal from Lake Okeechobee to the Gulf. It is part of the Inland Waterway System allowing large boats to cross Florida from the Atlantic to the Gulf. Since the canal is straight, little erosion occurs as it might along a normal river, and the water level is controlled. This is very different from the Peace River, which is famous for containing shark teeth or many of the other rivers in the state, which contain interesting fossil material. There, natural flooding, bank erosion, and run-off from tributary streams regularly expose new fossils. On the Caloosahatchee very little erosion takes place as it would on a normal river or from wave action on a lake. The banks in most places are overgrown and few fossils can be found.

When the canal was dug, large piles of spoil were deposited on the bank a short distance back from the river. In several places these spoil piles have been quarried for sand. In these abandoned quarries, some excellent fossils can be found. This sand, with its mixture of shells, is used for fill in subdivision and highway construction in the central part of the state. Therefore, looking at new construction sites and piles of highway fill often yields numerous fossil shells. It seems a shame that these shells are crushed and covered with asphalt along the highway, but there is no gravel in the central part of Florida.

If you are lucky enough to get permission to go into a shell pit near the river in the central part of the state, quite a few beautiful shells can be found. Over the years, we have been fortunate enough to make contacts with some shell pit operators and were able to convince them that we were more interested in looking for fossils than in vandalizing their equipment. Thus, we gained permission to look for fossils when the employees were not working. In this way we have acquired many fossils, which are now part of numerous collections, ours and others.

Many people ask us how we know that these shells are actually fossils since, in many cases, they are identical to present-day living shells found on the beaches along the coast. The most obvious answer is that these are saltwater forms, like barnacles, which are found well inland, approximately in the middle of the state. Some of these fossils are found in the position and orientation that they would have been when they were alive several thousands to a million years ago. None show any abrasion or other indications that they had been transported very far from where they originally lived. Therefore, these fossils were living in that location when saltwater covered most, or all, of central Florida. Many of these fossils can be identified in numerous and varied publications. Although hundreds of species have been identified, there is always the chance of finding something new to science. It is what keeps us going back!

Lee Creek Mine Pleistocene Fossils

A variety of Pleistocene marine mollusks are found in a large phosphate mine in North Carolina. This mine has produced a variety of well-preserved fossils, which have gone into the collecting community. Many of the **Pleistocene** fossils found here are similar or identical to those of the Caloosahatchee Formation of Florida.

Here is a continued selection of Pleistocene gastropods, mostly from Florida.

Olive (Ispidula) sayena: Caloosahatchee River. (Value range G for group).

Coralliophila mansfieldi: Caloosahatchee River. (Value range F for group).

Diodora floridana: Limpets are cap-shaped gastropods. Collectors of modern shells call these "coolie hats;" superficially they can resemble primitive, Cambrian mollusks known as monoplacophorans. (Value range G for group).

Cancellaria conradiana. (Value range G for group).

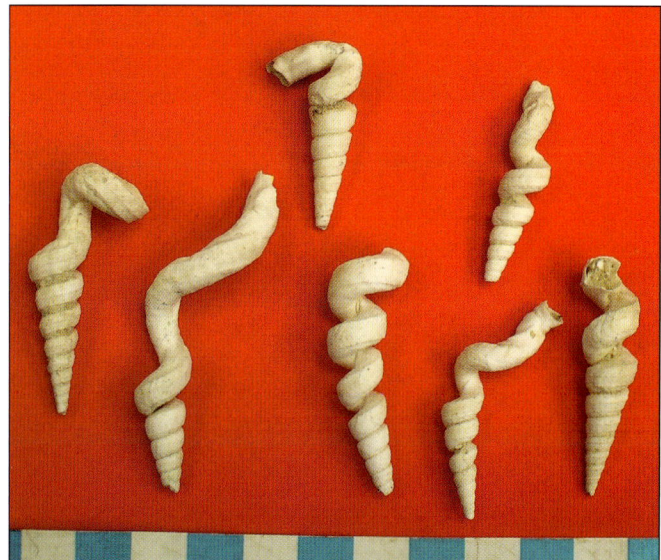

Vermicularia spirata: Another type (species) of this peculiar "worm-tube-like" gastropod. (Value range F for group).

Vasum horridum: Clewiston shell pit, Hendry Co., Florida. (Value range H).

Chicooreus floridanus: (very ornamented). (Value range G for group).

A single specimen of *Chicoreus* sp. (Value range H).

Turbo rhectogrammicus. (Value range G for group).

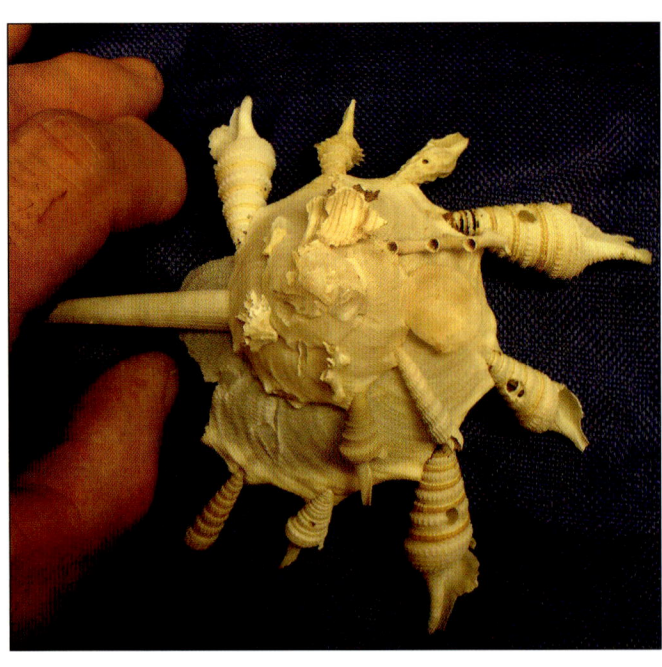

Xenophora sp.: Modern Xenophora gastropod shell with "foreign" shells attached—this mollusk does this naturally.

Xenophora conchyliophora: A gastropod that cements shell fragments and other "foreign" (viz. xeno.) materials in the construction of its shell. (Value range G for group).

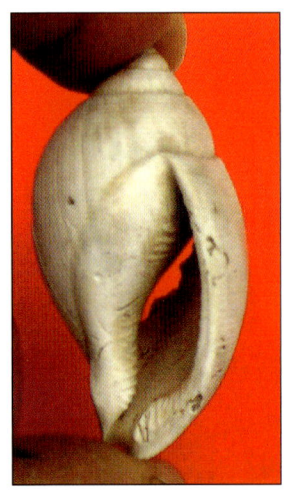

Sconsia hodgil. (Value range H).

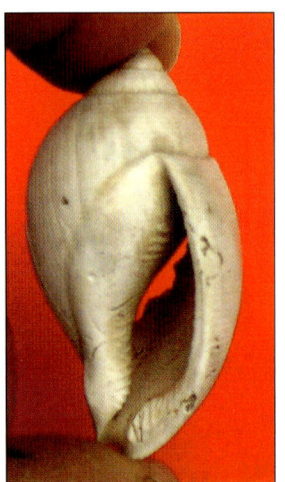

Polinices duplicatus: (fat and sassy!). (Value range G).

Natica plicatella: (plump—but a fine snail). (Value range G).

Trivia pediculus: (odd ornamentation). (Value range G for group).

Busycon sp.: lightning whelk, (an attractive genus—bold and full). (Value range H).

Fascicularia scalarina. (Value range G).

Strombus sarasetaensis. (Value range G, both specimens).

Petaloconchus floridana: Somewhat "mixed up" snails! (Value range G).

A smaller version of a "mixed up," sedentary gastropod like *Vermicularia*. Caloosahatchee Formation. (Value range G).

Crepidula sp. slipper shell has a brood chamber inside the shell. Caloosahatchee Formation, S. Florida. (Value range G for group).

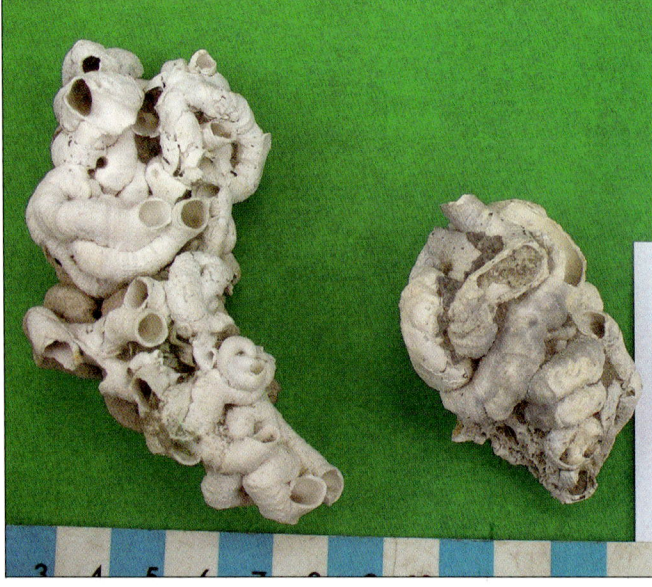

"Tube worm" gastropod: Yorktown Formation, Pliocene, Lee Creek Mine, North Carolina. (Value range G).

Gastropod, Pleistocene, S. California—Terrace deposits. (Value range H).

Group of gastropods from the Yorktown Formation, Lee Creek Mine, North Carolina: Note that these are quite similar to gastropods of the Caloosahatchee Formation of Florida. (Value range F for group).

Turritella pedroensis: Pleistocene terrace deposit, Los Angeles County, California. San Pedro sand. (Value range G).

Marine Pelecypods (Bivalves)

Next to gastropods, pelecypods (bivalves) are the most abundant mollusks in Pleistocene marine rocks.

Group of Caloosahatchee Formation mollusks, including pelycepods, cemented together. Caloosahatchee River, central Florida. (Value range G).

Platydon cancellatus: A nice Pleistocene marine clam from California. These clams come from Pleistocene terrace deposits, marls that occur along the west coast of North and South America and which formed when either the sea level was higher than it is today or the land in the area of the terrace deposit has uplifted since the mid-Pleistocene. (Value range H).

Dentalium nediexaganum: Scaphopods or tusk shells represent a minor molluscan phylum that, like most molluscan phyla, has a long geologic range. Late Pleistocene, San Pedro, California. (Value range G).

Chesapectin jeffersonius: Croatian Formation, Beaufort, North Carolina. Here is one of many species of this scallop shell. (Value range G for all specimens).

Chesapectin sp.: A group of some nice (and pretty) fossil shells. (Value range F for all).

Loess

When continental glaciers moved over the land, they produced great amounts of finely pulverized rock as they scoured underlying bedrock. This pulverized rock (rock flour) was carried by glacial melt-water into streams where it was then carried away. These numerous smaller streams, converging into large rivers like the Mississippi, the Missouri or the Yukon, became choked with this rock flour. Concurrent with the receding glaciers were prolonged winds. These winds **redeposited** this fine sediment into the uplands adjacent to these large streams. This windblown rock flour (or silt) is known as loess. It can contain fossils, most commonly those of land snails, but it can also contain the bones of mammoths and other elements of the Pleistocene megafauna—some of which may be from animals overwhelmed by these dust clouds.

Fossil Shells in Loess

During the Pleistocene, after each glacial advance, melt-water from the glaciers carried vast quantities of rock flour (pulverized rock formed from the movement of a glacier over bedrock) to lower elevations. This rock flour ladened water ultimately emptied into major rivers like the Missouri, Ohio or Mississippi, where it was deposited as immense silt bars. Recession of each glacier was accompanied by profound climatic change (warming), which brought on intense winds, these winds acting over many tens or hundreds of years. These winds picked up the rock flour along the rivers and redeposited it in **uplands** along the rivers. Such wind deposited silt, which was predominantly composed of rock flour derived from the glaciers, is known as loess (pronounced lurss—usually with an umlaut from the original German word).

This is an outcrop of loess along a stream in the St. Louis, Missouri, area. Natural outcrops of loess along streams often stand as a vertical bank, usually lack distinct bedding, and are quite homogeneous.

Loess outcrop: some distinct layers can be seen here in this natural outcrop, which is generally an uncommon phenomenon.

Helix sp.: Pulmonate gastropods with original coloration preserved. Mid-Pleistocene loess, Edwardsville, Illinois. (Value range G for group).

Terraced loess outcrop along a highway cut in Alaska east of Fairbanks. The loess here is associated with rock flour generated from continental glaciers whose silt laden melt water was washed into the Yukon River just to the north.

Left: *Helix* sp., Right: *Physia* sp.: Common land gastropods found in loess. This is a group of pulmonate gastropods (land snails) from Pleistocene loess of the St. Louis, Missouri, area. These are associated with the rock flour that "choked" the Mississippi River, a river that was in existence well before the Pleistocene Epoch. Similar fossil snails are found in the loess deposited along the Rhine River of Central Europe as well as at other places where loess occurs. (Value range H for group).

Pulmonate (air breathing) land snails from Pleistocene loess. Edwardsville, Illinois. (Value range G for group).

Succinea grosvenori: Valparaiso Sand, Wisconsin Glacial Stage, Shawneetown, Gallatin Co., Illinois. These high-spired shells are one of the most common land snails found in the loess. (Value range G for group).

Tropical Land Snails in Red Limestone of the Caribbean

The islands of the Caribbean saw sea levels drop when continental glaciers accumulated on the continents—water forming the continental glaciers having had to come from somewhere. Glacial build up lowered sea level and a cooler climate changed vegetation, allowing red, residual clay derived from upland regions to become incorporated into or to be derived from the weathered limestone—limestone which was now well above sea level. At this time, shells of large land snails were carried down along with red clay from the uplands and both incorporated into the broken and brecciated (by weathering) limestone and mixed with the limy sediment. On sea cliff and coastal outcrops of Pleistocene limestone in the Caribbean, these weathered (and brecciated) reddish zones of limestone, sometimes containing the shells of large land snails that lived (and live today) in the upland regions, can be seen.

Pupilla muscorum: A small gastropod found in loess that resembles the pupa of an insect—hence the name of the genus. Valparaiso Sand, Wisconsin Glacial Stage, Shawneetown, Gallatin Co., Illinois (Value range G, group of specimens).

A weathered, brecciated horizon in Pleistocene limestone—western Jamaica. During the build-up of ice on the continents, ice associated with periods of glaciation, sea level dropped. Limestone deposited during the previous interglacial stage (when sea level was at its highest) was weathered by exposure to erosion, weathering, and solution. This weathering was accompanied by red clay accumulation (residual clay originally incorporated in the limestone) or from clay being brought down from higher elevations and incorporated with the weathering limestone. Today, these red clay zones, representing periods of glacial build-up and a greatly lowered sea level, stand out conspicuously from the white Pleistocene limestone as can be seen in this outcrop in western Jamaica.

Presumably the uplands along these major rivers were rather sparse in life as loess, at least at most places, contains few fossils. Fossils that are found are usually the shells of land snails (pulmonate gastropods) along with the occasional bones and tusks of mammoths, sloth, and peccary. Loess is found over many parts of the Northern Hemisphere where it is usually a powdery, silty subsoil that is very homogeneous in appearance.

Breccia formed from partial solution of Pleistocene limestone during a period of **lowered sea level** caused as a consequence by the build up of continental glaciers which formed over large parts of the northern hemisphere. Lucia, western Jamaica.

Masonry wall made up of red-clay-containing Pleistocene limestone gathered from zones in Pleistocene limestones of the eastern Dominican Republic. This red rock was preferentially selected for its color in constructing this wall. The black rocks that floor the driveway on which the motorcycle sits are deep-sea cherts derived from the interior of the island.

Solution cavity filled with limestone breccia and red calcareous clay that forms the matrix of the breccia. This was formed from red, residual clay, which may have been washed down from upland areas when the local climate became drier as a consequence of glacial ice buildup on the continents. Washed down with the red clay were also shells of these large gastropods that are now found embedded in a brick red, calcareous matrix.

Limestone breccia broken up in the collection of these freshwater gastropods. Eastern Dominican Republic.

Large fossil pulmonate gastropods: These fossils were collected from the red clay rich limestone illustrated in the previous photos. These limestone breccias formed during periods of weathering which took place during periods of lowered sea level brought on by build-up of continental glaciers. The land snails may have been carried along with the red clay washed down from higher elevations or hermit crabs, which utilize the shells for protection against predators, may have carried them. They are the same species as live on the islands today, a frequent situation with Pleistocene invertebrates. Lucia, western Jamaica. (Value range F, single specimen)

Molluscan Fossils From a Pleistocene Swamp Deposit

During parts of the Pleistocene Epoch, glacial sediments deposited by the melted (or melting) glaciers would dam up streams, producing wet lands in the form of swamps. These swamps were a favored abode for animals of the Pleistocene megafauna, which included mastodons, ground sloths, and giant beavers. The most commonly found fossils in these swamp deposits are, however, those of fresh water mollusks, often of types that today live at higher latitudes. Associated with these cold water mollusks can also be found fossil plants, especially those of trees characteristic of cold climates such as spruce.

Freshwater clams and gastropods from Late Pleistocene swamp deposits. Note the blue speck below the snail; this is a "bleb" of vivianite, a ferrous phosphate mineral associated with Pleistocene bone rich sediments. (Value range G)

Late Pleistocene swamp deposits, St. Louis Co., Missouri.

Freshwater mollusks (gastropods and pelecypods) from Pleistocene swamp deposits, eastern Missouri. (Value range G).

Close-up of snail bearing clay: Elements of the Pleistocene megafauna also occur with these freshwater mollusk-rich clays.

Mollusk-rich clay with carbonized spruce cone. Late Pleistocene, eastern Missouri. (Value range G).

Small gastropods: Late Pleistocene swamp deposits, eastern Missouri. (Value range G).

Mercenaria sp.: Crystal clams! A large number of these clams, the centers of which have acted as geodes and become lined with honey yellow calcite crystals, come from the Fort Drum Crystal Mine in Okeechobee County, Florida. The beds in which they occur are of the same age as the Caloosahatchee Formation, which is the lower Pleistocene. These beds also yield a variety of gastropods and bivalves other than *Mercenaria*. (Value range G, complete single specimen).

Unio sp.: A river clam from late Pleistocene swamp deposits, eastern Missouri. Fossil river clams like this are not as common as one might think. They are very fragile, particularly in younger rocks, and generally fragment into small pieces when collected. (Value range F).

Bibliography

Brayfield, Lelia and William, 1986. *A Guide to Identifying Florida Fossil Shells and Other Invertebrates.* Self published. ISBN: 1-883167-02-7.

Brown, Robin C., 1988. *Florida's Fossils. Guide to Location, Identification, and Enjoyment.* Pineapple Press, Sarasota, Florida. ISBN 0-910923-4345-0.

Gould, Stephen J. 1985. "Opus 100 in The Flamingos Smile, Reflections" *in Natural History*. W. W. Norton and Co., New York-London. ISBN 0-393-30375-6. 1977.

_____. "The Misnamed, Mistreated and Misunderstood Irish Elk" *in Ever Since Darwin, Reflections in Natural History.* W. W. Norton and Co., New York and London. ISBN 0-393-00917-3.

Lange, Ian M., 2002. Ice age *Mammals of North America—A Guide to the Big, the Hairy and the Bizarre.* Mountain Press Publishing Co., Missoula, Montana. ISBN 0-87842-403-2.

Peterson, Carol and Bernie, 2008. *SOUTHERN FLORIDA'S FOSSIL SEASHELLS.* Self published. ISBN: 1-878398-69-5

Petuch, Edward J., 1998. *Geology and Paleontology of the Fort Drum Crystal Mine.* Okeechobee County, Florida. Edwin Rucks, Jr.—"publisher."

Chapter Nine
Pleistocene Insects

Insects in Copalite

Fossils depicted in this chapter are different from those normally seen; among their many attributes is that they are very attractive—venerable eye candy! The resins exuded from tropical trees since the Cretaceous Period have offered a medium which, upon hardening or polymerizing, are capable of preserving insects and other small organisms over long periods of time—virtually sealing them in a transparent sarcophagus. When the resin is exuded from a tree, it is in the form of a viscous, semi-liquid material in which insects, spiders, and even small reptiles can get stuck. It eventually hardens into a solid, clear, durable material. Such fossilized resin, generally referred to as amber, has been prized and utilized in commerce, usually as a gemstone, for over a thousand years. Amber, which is hard and stone-like, first became known from the Baltic region of northern Europe over 1,000 years ago. More recently it has been found in other parts of the globe and a number of various types of fossil resins also occur worldwide. These are referred to under various names, like Burmite from Burma, Kauri-gum from New Zealand, and copal (or copalite if the material is hard). Copal and Kauri-gum, in some cases, can be relatively soft and young, sometimes having been exuded just a few years ago. Copal is sometimes associated with still-living trees in the tropics, trees which may be growing upon soil which also contains copal; but copal which might be thousands, tens of thousands or hundreds of thousands of years old also occurs. A term which describes this hard, copal-related sub-fossil or fossil resin found occurring in the soil underlying copal producing trees is copalite (has the *ite* ending indicating that it comes from the earth). Copalite can be relatively hard and, like amber, sometimes is quite clear. It may also contain insect inclusions that can be readily seen through its clear composition, especially when polished.

Copal and copalite come from tropical trees, specifically the araucaracea, which under tropical or subtropical conditions can generate large amounts of resin which exudes from them—large amounts of resin which can be produced especially when the tree is injured or provoked, as it can be when harboring a nest of termites. The resin accumulates at the base of the tree on the forest floor, usually flowing, and in the process trapping various small life forms. Generally these are insects and spiders, but more rarely the resin can include small amphibians such as frogs and toads as well as small reptiles like lizards. Soft copal itself is a relatively modern resin, which accumulates from the resin producing trees and in the past has been dug or mined and utilized as a raw material in varnish making. Also found associated with the modern copal producing trees are older layers of resin accumulation which, unlike the modern or recent copal, can be hard and amber-like. The point at which copal becomes copalite is unclear and arbitrary. The rule of thumb is: if it's hard and brittle, it's copalite; if it's soft and semi flexible, it's copal. A graduation in hardness also exists between copalite and amber. Amber is harder and thus more cross-linked (polymerized) in its chemical make-up than is copalite. All of these transitions represent a gradational process involving the loss of volatiles as well as the gradual polymerization and cross-linking of the resin molecules. The lower (and hence older) layers of resin bearing soil can be the copalite of commerce, some of which can be amber-like and contain striking insect inclusions, while living copal producing trees may occur at the surface of such a deposit.

Determination of the age of copalite bearing layers can be difficult; ages can range from a few hundreds to a million years or more. Ages over 10,000 years qualify for a resin to be considered as a true fossil resin—and any organismic inclusions in it are then ***true fossils***. Copalite, with fossil inclusions (like those illustrated here), comes from three regions—eastern Africa, Colombia, South America, and Madagascar. Of the specimens shown, the copalite from east Africa is probably the youngest. East African copal was, until recently, mined and extensively used in varnish making, a use that has generally been supplanted by manmade, synthetic resins. East African copalite is also most affected by solvents, copalite from elsewhere usually being less so. A reasonable assumption can be made that the harder the resin and the less it is affected by a solvent (such as alcohol), the older it is. Colombian copalite is usually less affected by solvents. Madagascar copalite varies, some is affected, and some is not. This variance may reflect a variance in the age of the different deposits from which the copalite was obtained. The approach used here is to place all copalite under the Pleistocene, which can vary in age from 10,000 years (minimum) to over two million years. It might be mentioned that an age of Miocene or Pliocene has been stated for Colombian copalite; however, this appears to be influenced by dealers who want to give their wares the respectability of geologic age. Some dealers in fossils and minerals also label the Colombian copalite as amber, which probably is wrong as amber is harder and is not affected by solvents—but again, differences between amber, copalite, and copal are gradational and somewhat arbitrary. It should also be mentioned that all fossil resins are prone to craze with time. Crazing is the development of small cracks on a polished surface of the resin. Geologically young copalite is more prone to crazing than is amber. Copal is very prone to crazing as the volatile components in it evaporate. The degree of crazing a fossil resin undergoes in a decade or so may well be a function of its geologic age.

With regard to Colombian copalite—the source of many of the large masses containing spectacular inclusions—a reliable source of information on this material (The late Allan Graffham of Geological Enterprises) stated that it came from clay layers tectonically deformed, as shown in the initial photos of this chapter taken by him at one of the collecting sites. Strata involved in tectonic disturbance (tilted and/or faulted strata) have to be of some geologic age, a few hundreds of thousands of years at least, to be thus disturbed. Colombian copalite occurrences also appear to have different layers and/or areas of copalite. These different areas are of different geologic ages (a similar situation exists in Madagascar).

The matter of geologic age of the various fossil resins might be resolved by doing carbon-14 age dates on these various specimens. The problem with this is that C-14 tests are expensive and copalite from the same place may also differ in age depending upon where it was collected in the different copalite bearing layers. Copal producing trees in some tropical regions exist in the same place for hundreds of thousands of years, the resins exuded from them being buried in successive soil layers and varying considerably in age from top to bottom. With controlled stratigraphic collecting and careful documentation of where each piece was collected, such dating might give useful information. Such stratigraphic control, however, would currently be quite difficult to acquire at some of the producing areas, especially in Colombia as some are currently politically unstable and controlled by drug cartels.

Outcrop of copalite bearing clay beds, Santander Province, Colombia S. A. *Photo courtesy of the late Allen Graffham of Geological Enterprises, Ardmore, Oklahoma.*

The previous photo shows dipping (tilted) layers of copalite bearing strata; lines indicate direction of dip of strata. Picture taken c. 1983 and, as of this writing, Santander province is a region of revolutionary insurgents and is politically unstable and potentially dangerous territory; good pictures of these deposits are hard to come by.

All copalite specimens shown here are from Colombia, Santander province, unless otherwise indicated.

Group of copalite (amber?) specimens from Santander Province, Colombia, South America. There is a great deal of confusion as to what distinguishes amber from other fossil resins. The definition supported by the author is that **copal** is a soft and even gummy material currently exuded by living trees. **Copalite** by contrast is hard, having lost much of its volatile terpene components and its molecules being somewhat cross-linked (polymerized). Copalite may or may not be a true fossil material—the definition as to what is or is not a fossil having to do with age; if it is older than 10,000 years it is a fossil. Copalite generally can be considered as a fossil resin; amber being much older and more polymerized is definitely a fossil resin. A lot of copalite with insect inclusions is being sold on the fossil and rockhound market as amber. The author believes that this material should be sold as copalite and the term amber should be reserved for those fossil resins that are Miocene or older in age.

The same specimens as in the previous photo showing their original, unpolished surfaces.

Copalite slab containing winged termites. Allan Graffham of Geological Enterprises has referred to the Colombian fossil resin as amber—others refer to it inappropriately as copal. The author prefers to call it copalite (to indicate its hardness) in contrast to copal, which can be soft and even gummy. The designation copalite also indicates its **stone-like** or **gem-like quality** with the *ite* ending. (Value range E).

Different views of the same winged termite bearing copalite slab. (Value range E).

Same slab as shown in the previous photo.

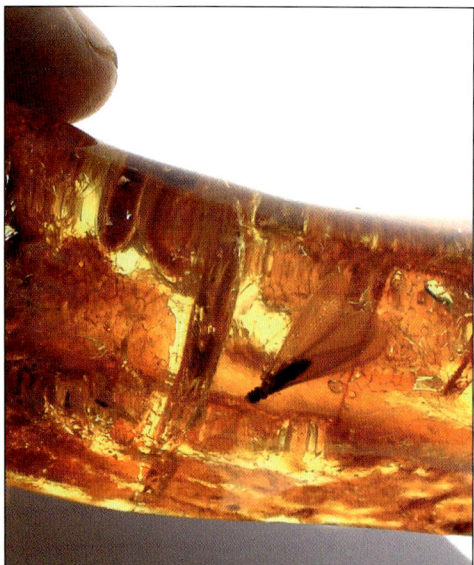
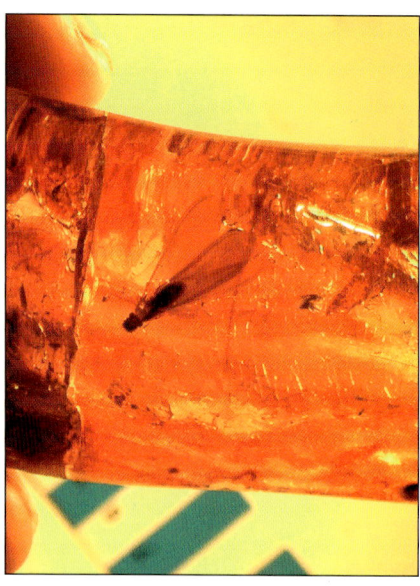

A single winged termite in a copalite chunk of smaller dimensions than that shown previously. Small worker termites can be seen to the left of the larger winged termites. In fossil resins there is a bias toward termites being preserved as activities by these insects can increase resin production and thus increasing the odds of termites being preserved in the resin. (Value range F).

View of the same specimen using different lighting to photograph through the transparent fossil resin.

Gracefully placed winged termite in a small slab of Colombian copalite. (Value range F).

Same specimen, different view: Note other inclusions in the slab and the granular appearance from the light going through the slab's rough back side.

Termite wings in an unusually clear copalite specimen. The circular sticker was placed by the original dealer to denote value of the specimen sold at the Tucson Show. (Value range E).

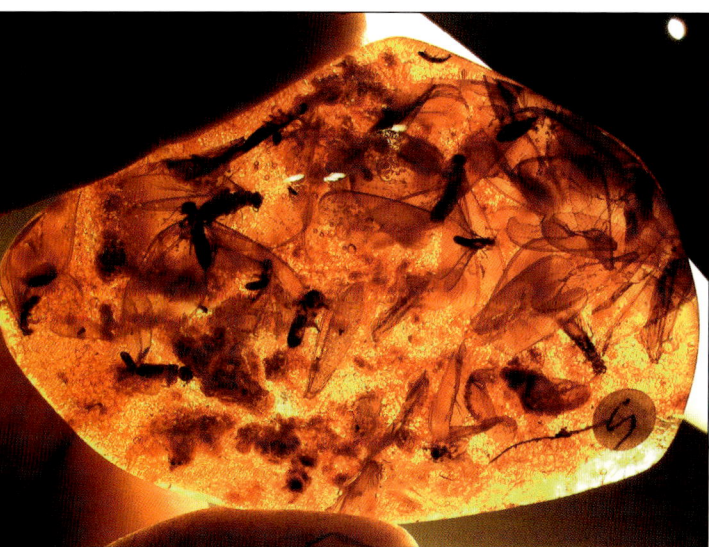

Large number of winged termites, most with the wings having become detached from the insects bodies by the flow of the resin when it was a viscous fluid. (Value range E).

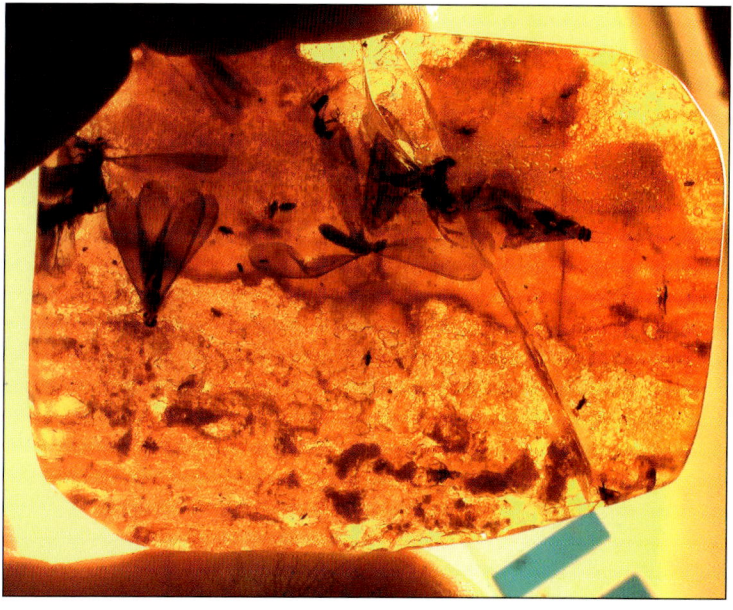

Complete winged termite at left: This copalite specimen sustains a fracture. Granular appearance is from light transmitted through the unpolished back of the specimen.

Frass: These elongate inclusions are wood fragments produced by the chewing of termites and known as frass. Small flies can be seen at the bottom of this very clear piece. All photos shown here represent specimens in which the insect inclusions can be observed with the unaided eye—a plus for these spectacular inclusions in copalite. (Value range F).

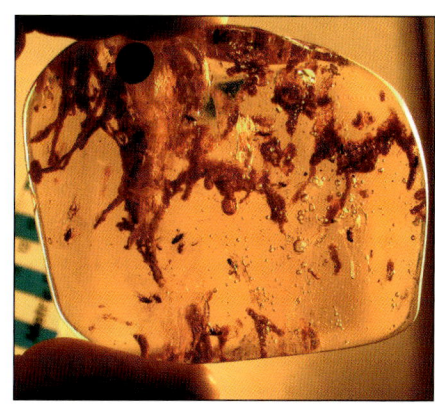
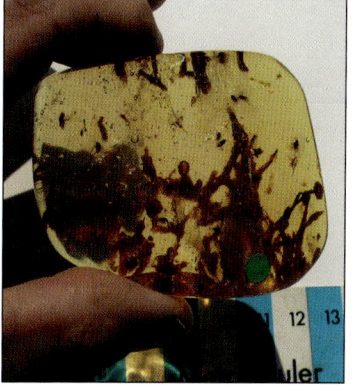

Same specimen as shown above with frass, but the photo was taken with a different lighting position.

Note the fragmented termites along flow lines of the resin. The termites were disarticulated and distended by flow prior to hardening of the resin. (Value range F).

Same specimen as above—close up of the winged termites—bubbles also obvious.

Large piece of copalite with two winged termites: Numerous bubbles in the copalite can be seen in this slab. (Value range F).

Two worker termites surrounded by frass, including some elongate strings of frass—lots of bubbles also. (Value range F).

Same specimen as above—different position and lighting.

Numerous worker termites and frass inclusions in Colombian copalite (Value range F).

Gnats. Numerous specimens of these small, pesky insects lovingly locked up (and no longer pesky) in clear copalite. (Value range E).

Gnats—same specimen as above—back side of the slab.

Large piece of copalite containing the usual winged termite fragments—with a (long hind leg) cricket at the bottom right. (Value range E).

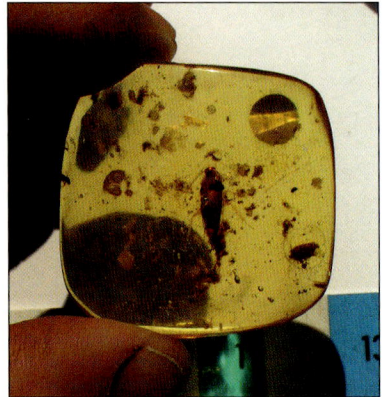

Cockroach (Blattoidea): A roach in fossil resin is a delight! It makes one think more favorably about these insects, which have been around since the Late Paleozoic when some were nearly a foot long. Roaches in this light are appealing and actually attractive when sealed in a fossil resin. Colombian copalite, Santander Province, Colombia. (Value Range E).

Close-up of cricket at the bottom right of the previously shown copalite chunk

Another roach—but this one in copalite that came from Madagascar. Madagascar copalite is similar to that from Colombia. It started to show up on the fossil market around 1999. (Value range F).

Another view of the previously shown cricket: Note the abundance of air bubbles in the copalite—ancient air?

Another view of the previously shown specimen under different lighting conditions showing the roach plus numerous smaller insects. Madagascar copalite. (Value range F).

Another roach in Madagascar copalite: Note layering in the resin; a consequence of flowage when the material was a viscous liquid. (Value range F).

A different view of the crane fly shown in the previous photo preserved in Colombian copalite.

Crane fly (family Tipulidae): A spectacular insect preserved in reddish Colombian copalite. (Value range D).

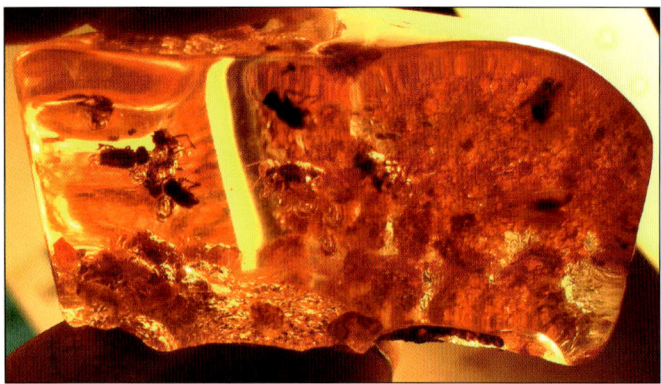

Twig boring beetles (family Bastrichidae) in Colombian copalite. (Value range E).

Twig boring beetles similar to the above specimen—Colombian copalite. (Value range E).

Gnats or small flies preserved in **Madagascar** copalite. This specimen has been only slightly polished. (Value range F).

Mosquitoes and ants preserved in very clear Madagascar copalite. As is the case with roaches, mosquitoes are not usually desirable things. Here are a few mosquitoes that are—all nicely encapsulated in their transparent fossil-resin coffins. (Value range E).

Ants in clear Colombian copalite: This specimen was collected over 60 years ago and was part of the geologic collection of St. Louis University, which was given to the St. Louis Science Center. In 1995, the display case containing this specimen was broken into and this specimen stolen. *Photo courtesy of Jim Houser, former curator of the St. Louis Museum of Science.*

Slab of east African copalite with a lot of vegetable debris: Note crazing on the specimens surface, a property of copalite that has set for a few years after being polished.

Ants in clear **Madagascar** copalite: Madagascar copalite has been labeled as being Pliocene in age by some fossil dealers. Again, the age of this fossil resin is debatable and, as is the case with Colombian copalite, individual specimens may be representative of the different ages of the localities where it is obtained. (Value range F).

Another view and position of a copalite specimen crowded with large black ants (Eastern Africa copalite). This copalite may be the youngest shown here—possibly Holocene in age rather than Pleistocene. Eastern Africa copal a few decades ago was mined and extensively used in the making of varnish; synthetic resins today serve the same purpose.

Numerous ants in eastern African copalite: Eastern Africa copalite (and copal) used to be mined and used in the making of varnish. This material might be Holocene in age, as it is not as hard as that from Colombia or Madagascar. (Value range F).

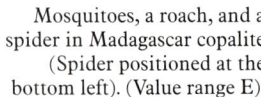
Mosquitoes, a roach, and a spider in Madagascar copalite (Spider positioned at the bottom left). (Value range E).

Close-up of the spider in the previous photo, with a fly above it. Madagascar copalite. Spiders are rarer in fossil resins than are insects.

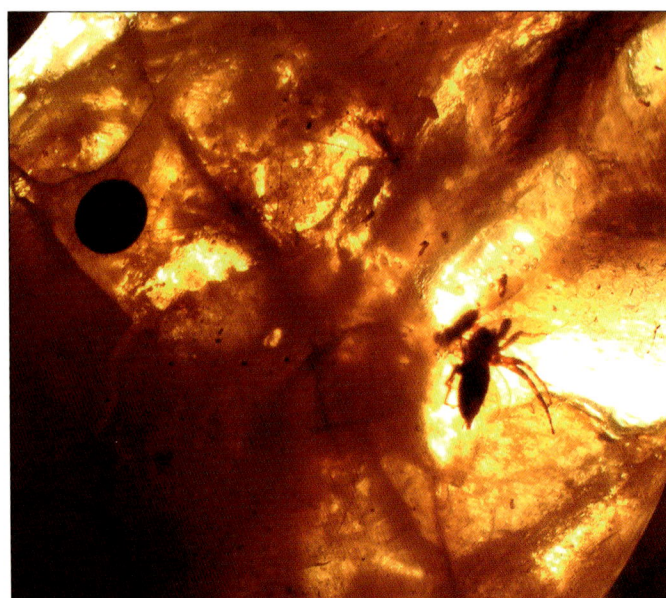
Spider in a large piece of Colombian copalite: The circle to the left of the spider is a colored tag used by the importer to classify (according to price) the value of the piece with its fossil inclusions. (Value range E).

A fairly large spider in Colombian copalite. (Value range E).

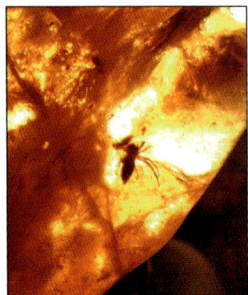
Enlarged view of the same spider.

Spider in copalite, enlarged view: Note the air bubbles in the piece at 10:00 from the spider. Amber can also contain such air bubbles, which in the case of true amber can be a small sample of the earth's atmosphere from millions of years ago.

Spider and small insects in Madagascar copalite (Value range F).

Spider in Madagascar copalite: Note detached leg torn by the flow of the resin when still a viscous liquid. (Value range F).

Leaf in Madagascar copalite. (Value range F).

These are real scorpions embedded in polyester resin to resemble fossils in amber or copalite—**they are fake fossils**! Such "bugs" in resin make interesting ornaments, but should not be fraudulently sold as the real thing, which is quite rare and therefore pricey.

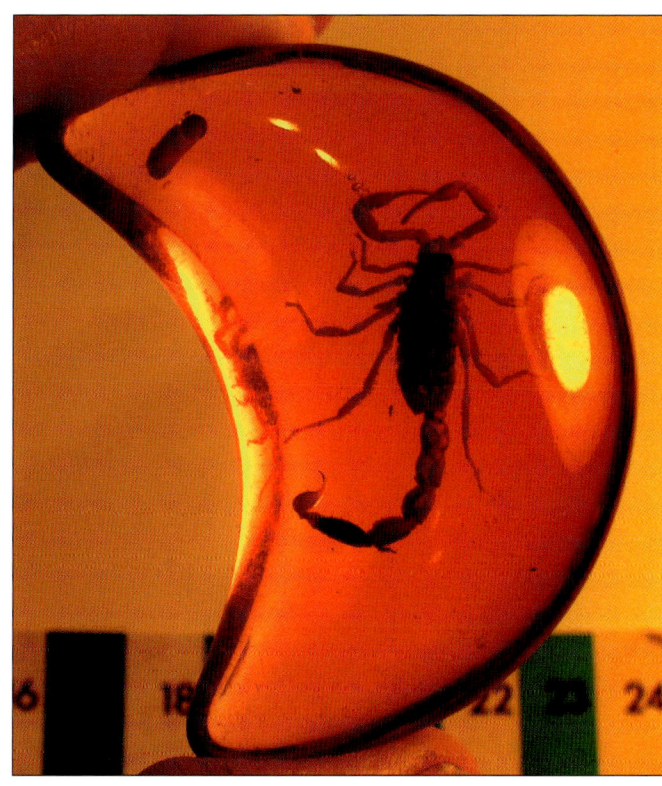

A real scorpion (but **fake** "fossil" resin) sold as an amulet. (Value range G, as an amulet, not as a fossil).

Insects Preserved in Asphalt

Pleistocene insects preserved in asphalt.

A group of scorpion inclusions in manmade resin.

Cybister (Hydrophilus) explavatus (partial): Predaceous water beetle, Rancho La Brea Formation, Kern County, California. It's been said in the Creation-Evolution controversy, "God must have had an inordinate fondness for beetles as he made so many different kinds of them" (presumably this would also include fossil ones). (Value range F).

Small arthropods embedded in polyester resin and then polished to resemble a "bug" in amber. Are these real arthropod inclusions being suggested as actual (fake) trilobites? As with many other seemingly spectacular "fossils" on the fossil market, it's Caveat Emptor.

Cybister (Hydrophilus) explavatus: These large beetles fed on carrion from the large mammals that died stuck in the tar pits.

Cybister (*Hydrophilus*) *explavatus*: A particularly nice and complete specimen in asphalt. Rancho La Brea Formation, Kern County, California. (Value range E).

Scarab beetles collected from asphalt saturated soil zones, which occur sporadically in the Los Angeles basin of California and are sometimes uncovered during excavations. These isolated occurrences are distinct from the famous Rancho La Brea tar pits. These insects came from a mass of asphalt-saturated Pleistocene soil exposed while digging a water main. (Value range G, single specimen).

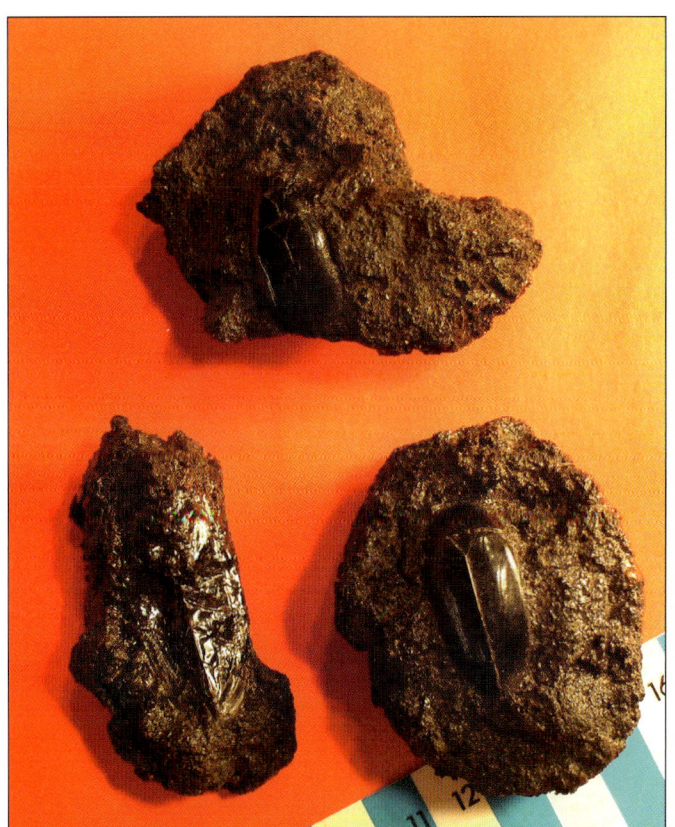

Group of Cybister beetles: Specimen at the bottom left shows dorsal (bottom) of the beetle.

Bibliography

Poinar, George O., 1992. *Life in Amber*. Stanford University Press. Stanford California.

_____. 1994. *The Quest for Life in Amber*. Helix books, Addison Wesley Publishing Co. ISBN 0-201-62660-8

_____ and Roberta Poinar, 1999. *The Amber Forest*. Princeton University Press, Princeton, New Jersey.

Ross, Andrew, 1999. *Amber*. Harvard University Press, Cambridge Mass., ISBN 0-674-01729-3.

Chapter Ten
Pleistocene Vertebrates

Introduction

The Pleistocene Epoch includes the "Great ice age," a time (or times) when large parts of the earth's land masses were covered by continental glaciers, especially in the Northern Hemisphere. The Pleistocene includes **four** major periods of glaciation that were sandwiched between periods when the climate was warmer than it is today—the interglacial stages. Cooler climates, associated with glacial ice buildup, resulted in entire ecosystems moving to **lower latitudes**. When the ice melted and warmer conditions prevailed, warmer climate ecosystems migrated northward again to **higher latitudes**. Thus, in any particular area, Pleistocene strata can have cold climate faunas alternating with those of a warmer climate. It was during the Pleistocene that humans migrated over the globe from their place of origin in Africa. The African continent, being "our birthplace," also has the best fossil record of humanity. Humans may possibly have been a consequence of the ice age itself, as Africa's climate changed drastically during the Pleistocene.

Pleistocene continental glaciation, both to the north and south of Africa, turned much of sub-Saharan Africa from a tropical climate to a more savanna-like one—a change possibly responsible for the evolution of human intelligence and cultural development. The Pleistocene, with its ice ages and glaciations, appears to have been a **unique part of geologic time**.

Pleistocene Extinctions

One of the major extinction events in the fossil record occurs at the end of the Pleistocene Epoch—especially with regard to large mammals. This extinction (known as the Pleistocene megafauna extinction) is generally attributed to the spread of humans around the globe toward the end of the ice age. This event is biased toward large animals (the megafauna); smaller animals appear to have been largely unaffected by it.

Arguably, the Pleistocene is one of the most interesting parts of geologic time. With respect to collecting its fossils, Pleistocene fossils are also some of the most accessible. In many parts of the globe a creek, brook, river bank or gravel bar holds the potential for yielding animal bones and teeth of the megafauna and Pleistocene marine strata are often chuck full of fossils. To begin this chapter, a "guest essay" by Charles Lyell is provided, utilizing a short segment from his book, *Second Visit to the United States of America*, 1846-'47.

Lyell embarked on an ambitious trip in 1846 to observe what, at that time, was the western terminus of Anglo-America civilization. In this trip he utilized two new transportation modes, the fledging railroad and the steamboat, which took him across the southern states, where he observed (and commented upon) ferment that would erupt as the American Civil War a decade and half later. Going by steamboat up the Mississippi from New Orleans, he made keen observations on Pleistocene outcrops in river bluffs, particularly near Natchez and Vicksburg, Mississippi. In a large ravine near Natchez were found human remains (Natchez man) that appeared to have come from the Pleistocene loess. Lyell's ambivalence regarding the antiquity of Natchez man was appropriate and the issue of the antiquity of humanity in North America is still a disputed issue today, even with technological advances like Carbon-14 age dating. Indeed the matter of the appearance of "modern man" and his spreading over the globe transcends all other issues when it comes to the Pleistocene Epoch.

Natural outcrop of Pleistocene loess, a wind blown sediment associated with glaciation. Large gullies, eroded into this soft "rock" have produced elements of the ice age megafauna—particularly in the lower Mississippi Valley of Mississippi and Louisiana. This is a typical outcrop of this wind deposited material—notice how it forms steep vertical banks.

> At Natchez (where I rejoined my wife), there is a fine range of bluffs, several miles long, and more than 200 feet in perpendicular height, the base of which is washed by the river. The lower strata, laid open to view, consist of gravel and sand, destitute of organic remains, except some wood and silicified corals, and other fossils, which have been derived from older rocks; while the upper sixty feet are composed of yellow loam, presenting, as it wastes away, a vertical face towards the river. From the surface of this clayey precipice are seen, projecting in relief, the whitened and perfect shells of land-snails, of the genera *Helix, Helicina, Pupa, Cyclostoma, Achatina,* and *Succinea.* These shells, of which we collected twenty species, are all specifically identical with those now inhabiting the valley of the Mississippi.

Account by Charles Lyell, from his *A Second Visit to the United States of North America*, 1846-'47, where he discusses Pliocene (the lower strata) and Pleistocene loess exposed in Mississippi River bluffs near Natchez, Mississippi. Note that here is mentioned the ever-present fossil snails (*Helix, Pupa,* etc.) illustrated in chapter eight. The "silicified **corals** and other fossils," which Lyell mentions, was probably *Favosites* sp., a Paleozoic fossil in chert probably derived from Devonian rocks to the northeast in Tennessee.

> CHAP. XXXI.] OF NATCHEZ. 195
>
> The resemblance of this loam to that fluviatile silt of the valley of the Rhine, between Cologne and Basle, which is generally called "loess" and "lehm" in Alsace, is most perfect. In both countries the genera of shells are the same, and as, in the ancient alluvium of the Rhine, the loam sometimes passes into a lacustrine deposit containing shells of the genera *Lymnea, Planorbis,* and *Cyclas,* so I found at Washington, about seven miles inland, or eastward from Natchez, a similar passage of the American loam into a deposit evidently formed in a pond or lake. It consisted of marl containing shells of *Lymnea, Planorbis, Paludina, Physa,* and *Cyclas,* specifically agreeing with testacea now inhabiting the United States. With the land-shells before mentioned are found, at different depths in the loam, the remains of the mastodon ; and in clay, immediately under the loam, and above the sand and gravel, entire skeletons have been met with of the megalonyx, associated with the bones of the horse, bear, stag, ox, and other quadrupeds, for the most part, if not all, of extinct species. This great loamy formation, with terrestrial and fresh-water shells, extends horizontally for about twelve miles inland, or eastward from the river, forming a platform about 200 feet high above the great plain of the Mississippi. In consequence, however, of the incoherent and destructible nature of the sandy clay, every streamlet flowing over what must originally have been a level table-land, has cut out for itself, in its way to the Mississippi, a deep gully or ravine. This excavating process has, of late years, proceeded with accelerated

---

> 196 RAVINES IN TABLE-LAND. [CHAP. XXXI.
>
> speed, especially in the course of the last thirty or thirty-five years. Some attribute the increased erosive action to partial clearings of the native forest, a cause of which the power has been remarkably displayed, as before stated, within the last twenty years, in Georgia.* Others refer the change mainly to the effects of the great earthquake of New Madrid, in 1811—12, by which this region was much fissured, ponds being dried up and many landslips caused.
>
> In company with Dr. Dickeson and Colonel Wailes, I visited a narrow valley, hollowed out through the shelly loam recently named "the Mammoth ravine," from the fossils found there. Colonel Wiley, a proprietor of that part of the State of Mississippi, who knew the country well before the year 1812, assured me that this ravine, although now seven miles long, and in some parts sixty feet deep, with its numerous ramifications, has been entirely formed since the earthquake. He himself had ploughed some of the land exactly over one spot which the gully now traverses.
>
> A considerable sensation was recently caused in the public mind, both in America and Europe, by the announcement of the discovery of a fossil human bone, so associated with the remains of extinct quadrupeds, in "the Mammoth ravine," as to prove that man must have co-existed with the megalonyx and its contemporaries. Dr. Dickeson showed me the bone in question, admitted by all anatomists to be part of a human pelvis, and being a fragment of
>
> * See ante, p. 25.

Mentioned here are the large gullies or ravines eroded into loess. The author observed and collected fossils in the South in the late 1950s and early '60s from gullies similar to those mentioned by Lyell. These were formed in soft Cenozoic sediments, which included loess. Today most of these are overgrown. Is this vegetation a consequence of an increased amount of carbon dioxide associated with accelerated burning of fossil fuels and the greenhouse effect since that time? These ravines were (and are) a source of Cenozoic fossils, especially vertebrates. Page 196, 197 of Lyell contains an interesting account of a fossil. Human bone (Natchez man) was found in the "mammoth Ravine," one of these large gullies.

More of Lyell's discussion on fossil snails found in loess outcrops and then mention of some of the elements of the Pleistocene megafauna like mastodon and the giant ground sloth (*Megalonyx*).

> CHAP. XXXI.] FOSSIL HUMAN BONE. 197
>
> the *os innominatum*. He felt persuaded that it had been taken out of the clay underlying the loam, in the ravine above alluded to, about six miles from Natchez. I examined the perpendicular cliffs, which bound a part of this water-course, where the loam, unsolidified as it is, retains its verticality, and found land-shells in great numbers at the depth of about thirty feet from the top. I was informed that the fossil remains of the mammoth (a name commonly applied in the United States to the mastodon) had been obtained, together with the bones of some other extinct mammalia, from below these shells in the undermined cliff. I could not ascertain, however, that the human pelvis had been actually dug out in the presence of a geologist, or any practised observer, and its position unequivocally ascertained. Like most of the other fossils, it was, I believe, picked up in the bed of the stream, which would simply imply that it had been washed out of the cliffs. But the evidence of the antiquity of the bone depends entirely on the part of the precipice from which it was derived. It was stained black, as if buried in a peaty or vegetable soil, and may have been dislodged from some old Indian grave near the top, in which case it may only have been five, ten, or twenty centuries old; whereas, if it was really found in situ at the base of the precipice, its age would more probably exceed 100,000 years, as I shall endeavour to show in a subsequent chapter. Such a position, in fact, if well authenticated, would prove that man had lived in North America before the last great revolution in the physical geography of

The ambiguity presented here by Lyell regarding the antiquity of man in North America persists to the present day. Note that Lyell's age dates are relatively close to those given today for similar Pleistocene fossils—the modern dates having the advantage of carbon-14 age-dating.

> 198 TORNADO AT NATCHEZ. [CHAP. XXXI.
>
> this continent had been accomplished; in other words, that our race was more ancient than the modern valley, alluvial plain, and delta of the Mississippi,—nay, what is more, was antecedent to the bluffs of Port Hudson and Natchez, already described. Now that elevated freshwater formation, as I shall by and by endeavour to show, is the remnant of a river-plain and delta of extremely high antiquity; and it would follow, if the human race was equally ancient, that it co-existed with one group of terrestrial mammalia, and, having survived its extinction, had seen another group of quadrupeds succeed and replace it.

Hypothesis on the antiquity of man antecedent to the time of the formation of the Mississippi River valley.

Pleistocene glacial gravels: The presence of continental glaciers over a large part of the earth's continents—particularly in the northern hemisphere—caused a large amount of material to be scraped, transported, and re-deposited as glacial till (moraine), like these beds of gravel exposed along an Alaskan stream. These deposits are typical of the sediments representative of the Pleistocene Epoch of the Quaternary Period—the last and youngest subdivision of geologic time.

Deer

Deer bones and antlers can be common Pleistocene fossils. Pleistocene deer were pretty much the same as those of today.

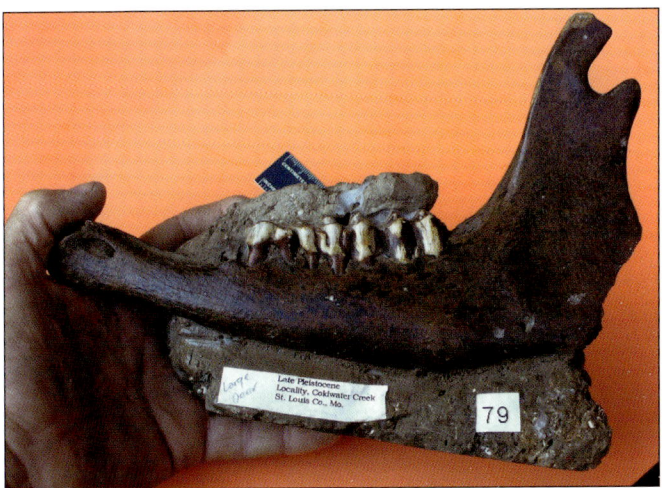

Deer jaw: Deer in the Pleistocene were essentially the same as those of today. Late Pleistocene, Rancholabreain, Coldwater Creek, eastern Missouri.

Caves like this along an Ozark stream can contain the bones of Holocene and Pleistocene animals. Such bones can be preserved either by being covered with dripstone like that forming the two slender columns or they can be preserved in the clay of the cave floor.

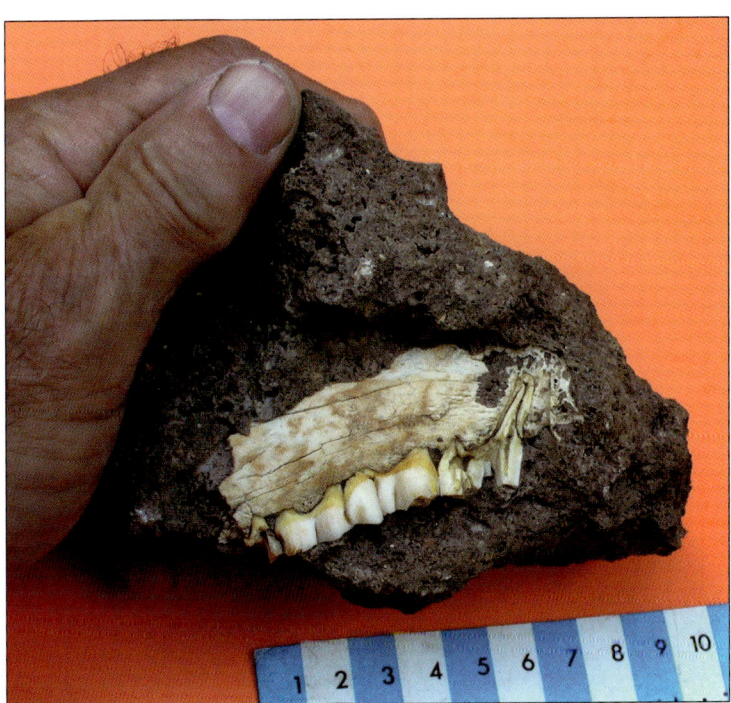

Deer jaw in travertine: Usually Pleistocene fossils are not found in hard rock; these bone and jaw fragments represent an exception. They are preserved in a type of limestone known as travertine, which was deposited from a spring. Travertine is often geologically young and is sometimes sliced and used as decorative stone. Travertine is also (in part) the secondary material deposited in caves and large cavities in limestone and dolomite; travertine and related dripstone can also preserve fossils in caves. (Value range F).

Pleistocene fossils are found in a variety of different geologic environments— the sediments derived from them often being of a greater variety than are those of most earlier parts of geologic time. These include loess, glacial, and river gravel deposits as well as caves and tar pools. Pleistocene fossils often consist of the bones of very large mammals representative of the **Pleistocene megafauna,** mammals which often were quite a bit larger than related ones living today.

Group of fossil deer bones in travertine: Deer from the Pleistocene are the same species as exists today. These specimens came from travertine exposed in an excavation for a subdivision in Jefferson Co., Missouri. (Value range F, single specimen).

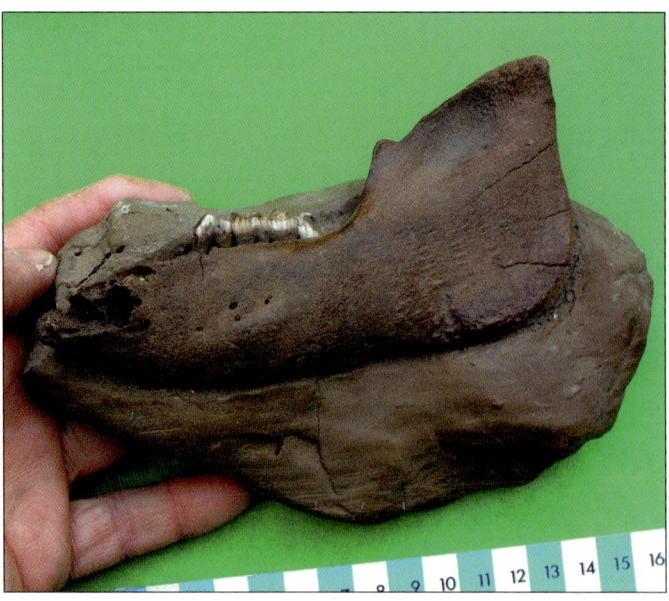

Peccary jaw: Peccary's were (and are) pig-like animals that currently live in the southwestern U.S. and in Mexico. During the Pleistocene, especially during the warmer interglacial periods of that epoch, large herds of peccaries lived in the U.S.—their bones locally can be abundant where members of these herds met an untimely death.

Mineralized deer antler: These can be somewhat common fossils where Pleistocene vertebrates are concentrated. Santa Fe River, Florida. (Value range G).

Casteroides sp.: Jaw section of a giant beaver. *Casteroides* is one of the large mammals constituting part of the Pleistocene megafauna—a fauna of particularly large mammals that went extinct at the end of the Pleistocene Epoch. Their extinction is suggested to be a consequence of the rapid spread of humans over the globe at that time. (Value range C).

Rodents

Large rodents form part of the Pleistocene megafauna. The giant beaver *Casteroides* sp. is a particularly notable example.

Casteroides sp.: Left: same specimen as in previous photo; Right: lower incisor in matrix. Late Pleistocene, Coldwater Creek, St. Louis Co., Missouri. (Value range B for all).

Casteroides sp.: Three specimens of incisors of this large beaver are shown on the left. Lower jaw of a modern beaver shown for comparison—it is to the right of the three specimens (light colored, modern bone); and to the far right is a lower jaw with incisor. Coldwater Creek, eastern Missouri. (Value range B for group).

Cast of a complete Casteroides skull. (Value range E).

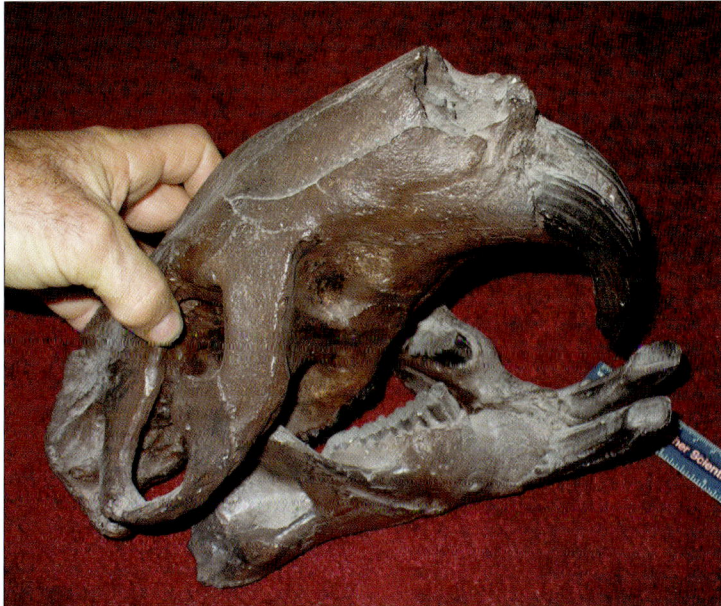

Another view of the same skull.

Edentates (Glyptodonts and Armadillos)

Edentates moved into the Northern Hemisphere from the southern during the Pleistocene when a land bridge in Central America developed between South and North America; armadillos survived, the Glyptodonts went extinct at the Pleistocene's end in both hemispheres.

Holmesina septentionalis: armadillo scute. These scutes, like those of Glyptodon, come from a large, extinct armadillo, which, like other edentates, originated in the Southern Hemisphere. Polk County, Florida. (Value range F).

Glyptotherium sp.: Glyptodon scutes. These large armadillo-like edentates lived to the end of the Pleistocene. In the U.S. their remains (most commonly scutes like these) are found in Florida as well as elsewhere in the southern states. Glyptodonts originated in the southern hemisphere where they migrated into North America during the Pleistocene when sea level dropped, exposing a land bridge in central America—this land bridge allowed their migration into the northern hemisphere. (Value range G, single scute).

No! They are not dinosaurs, however elements of the Pleistocene megafauna have found their way into kids attractions and toys, as many were *large*, *strange* and *extinct*. Here giant endentates take the stand.

Glyptotherium sp.: A Glyptodon scute similar to those shown above, but this one came from the Dallas-Fort Worth area, Texas.

Sloths

Large ground sloths were one of the distinctive elements of the megafauna.

Eremotherium rusconi: A tooth (cast) of the largest known ground sloth. These huge teeth are known from the Lower Pleistocene of the southern U.S.

Megalonyx claw shortly after being found by Rich Hagar. Claws of this large ground sloth are one of the most distinctive fossils of the Pleistocene megafauna. Finding a nice fossil like this is exciting and is somewhat like winning the lottery—and like gambling, it can be addictive to some people!

Megalonyx jeffersoni: *Megalonyx* means giant claw and was originally thought to be the claw of a giant carnivore. The description of this vertebrate fossil was essentially the beginning of vertebrate paleontology in North America. The genus *Megalonyx* was described in 1799 by Thomas Jefferson in "Transactions of the American Philosophical Society" under the title "A Memoir of certain bones of a Quadruped of the clawed kind in the western parts of Virginia." These are the "giant claws" of this interesting American fossil. Late Pleistocene, Coldwater Creek, eastern Missouri. (Value range C).

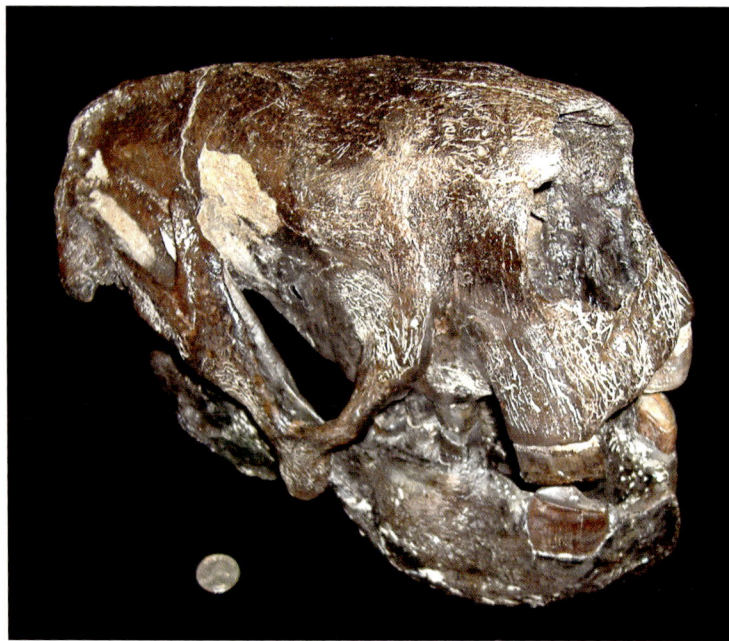

Megalonyx jeffersoni: This is the peculiarly shaped skull of this large ground sloth. The species *jeffersoni* was added when it was realized by Caspar Wistar, in the early 1800s, that what had been originally described by Jefferson as a carnivore was **not**, but rather represented an extinct type of sloth. Wistar named the species after Jefferson, who originally had established and described the genus *Megalonyx*. In the late Pleistocene *Megalonyx* appears to have been somewhat of a common denizen of the Midwestern U.S. as its remains can locally be abundant in this part of the country. This specimen is from the Kansas (Kaw) River in eastern Kansas. (Value range C).

Megalonyx jeffersoni: Besides the claws, these peg-like teeth are another distinctive feature of this large ground sloth. Late Pleistocene swamp deposits, Coldwater Creek, eastern Missouri.

A Pleistocene megafaunal element as a stuffed toy—a giant ground sloth.

Megalonyx jeffersoni: This is a relatively large, single tooth of Jefferson's ground sloth. Shown is the blunt grinding surface of the tooth. (Value range E).

Bison

A number of large, extinct bison species lived during the Pleistocene. The modern *Bison bison* is the only species to have survived these extinctions. Most Pleistocene bison species were larger than the modern *Bison*.

Bison antiquis: Another extinct species of bison from the gravels of the Missouri River. (Value range D).

Bison bison: This is one of the species of bison that lived during the ice age. *B. antiquis* is one of the more frequently found extinct bison species. All species of bison suffered extinction at the end of the Pleistocene with the exception of *Bison bison*, which became the food staple of the Plains Indians up to the late nineteenth century when their large populations were decimated by westward expansion. From sediments of the Missouri River upstream from St. Louis, Missouri. (Value range F).

Elk-Moose

The extinct genus *Cervalces* was a major part of the Pleistocene megafauna.

Bison antiquis teeth: Flint River, Georgia.

Cervalces scotti: This is the rack of a large elk (or stag moose), an element of the Pleistocene megafauna. The width across the rack is approximately eight feet. *Cervalces* is often referred to as the "stag moose" or the "elk moose," it seems to be somewhat of a combination between these two mammals. Kansas (Kaw) River, eastern Kansas. (Value range C).

Cervalces scotti (skull cap): The two conspicuous "knobs" on this skull are detachment areas for the rack of this stag-moose. Kansas (Kaw) River, eastern Kansas. (Value range D).

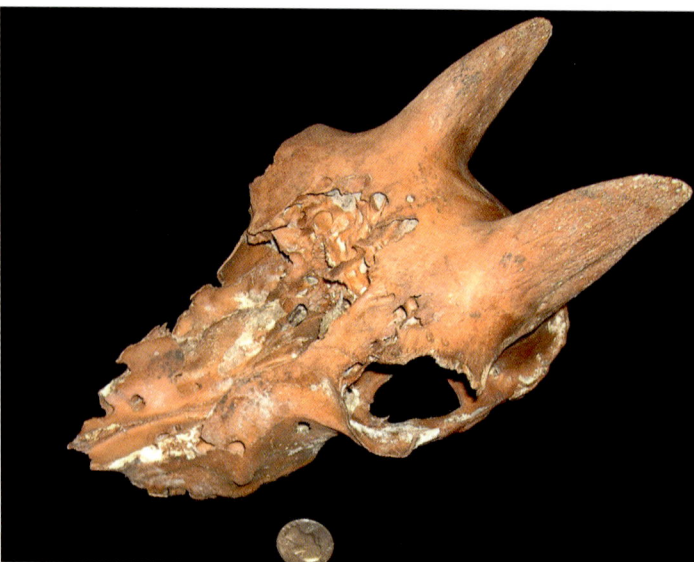

Goat: Placer gravels near Livengood, Alaska. Goats of the Pleistocene were similar to wild goats, like the mountain goats of today. (Value range E).

Camel and Horse

Widespread in North America and elsewhere during the Pleistocene was the camel and horse. Both animals went extinct at the end of the Pleistocene, but humans later reintroduced the horse onto the continent. Horse and camel teeth are fairly common Pleistocene fossils, but with the former it is often difficult to determine if those found are really from the Pleistocene (hence a native horse of North America) or whether they are from the modern (Holocene) horse, which was reintroduced by Europeans in the sixteenth century.

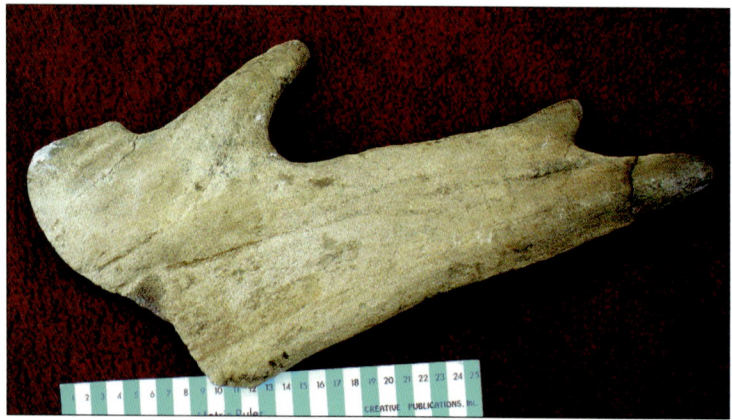

Cervalces scotti: Part of a palmate rack of the stag or elk moose. Kansas (Kaw) River sediments, eastern Kansas. (Value range F).

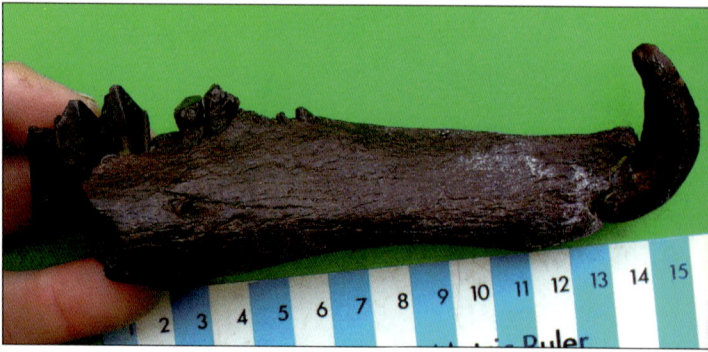

Camelops sp.: Jaw of a North American camel. Camels were part of the mammalian fauna of North America until they became extinct on the continent at the end of the Pleistocene. Santa Fe River, Florida. (Value range F).

Equis sp.: Horse teeth. Horses were dominant large herbivores in North America during the Pleistocene. Like other representatives of the Pleistocene megafauna, they went extinct at the end of the epoch—some 10,000 years ago. Europeans reintroduced them into the continent in the sixteenth century. (Value range F for group).

Tapir jaw: Suwannee River, northern Florida. (Value range F).

Tapir

The tapir lived over a large portion of North America during the Pleistocene, especially during the colder periods of glacial buildup. Today tapirs are associated with high latitudes.

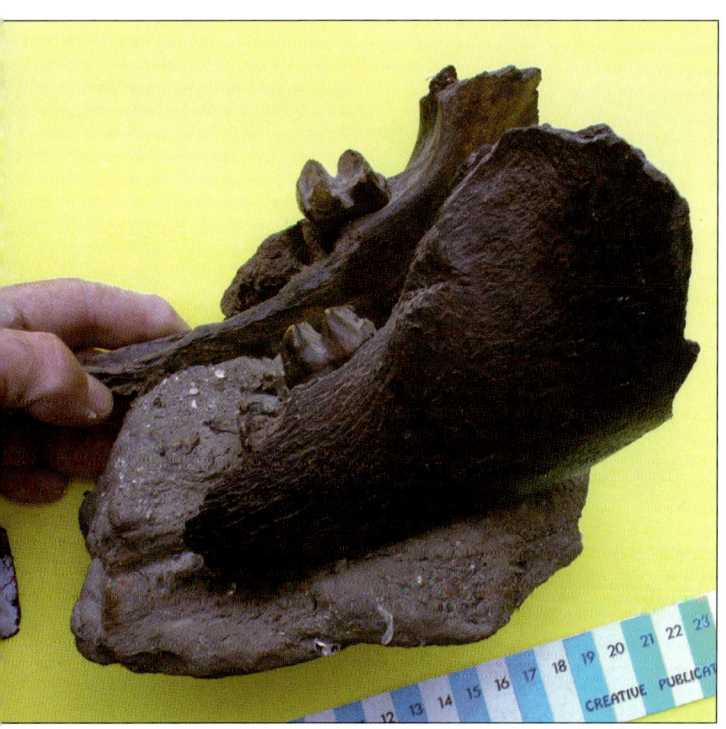

Tapir: Suwannee River, northern Florida. The tapir today lives only at high latitudes. (Value range E, reconstructed).

Mammuthus primigenius. Large teeth of the "wooly mammoth" can be relatively common Pleistocene fossils. This probocidians tooth-grinding-surface is at the bottom right. Teeth of the wooly mammoth (*M. primigenius*) are often preserved (in part) as a consequence of their large size and their toughness. In addition to this toughness, their uniqueness of shape aids in their being recognized as a peculiar object and collected by persons who otherwise would have no knowledge of interest in fossils. These teeth, because they are often found surreptitiously, sometimes show up in flea markets and jumble shops. Mid Pleistocene, St. Charles County, Missouri. (Value range C).

Mammoths

Mammoths are probably the best known of the Pleistocene megafauna. Fossils of mammoths, particularly their teeth, are especially recognizable and desirable fossils. Mammoths were closely related to elephants; they were essentially "hairy elephants."

M. colombi: A different view of the water worn and manganese dioxide stained tooth shown above from the Cooper River, South Carolina.

Mammuthus cf. (probably) *M. colombi*: A medium-sized mammoth tooth found in cave sediments of the eastern Ozarks. The tooth is covered with a thick layer of manganese dioxide, a mineral commonly found associated with ancient objects found in the streams flowing through caves. (Value range D).

Mammuthus colombi: The grinding surface (somewhat worn) of a medium aged mammoth. Teeth from mammoths that reached old age generally show severe wear, as their diet was believed to have been grass, a plant that can be highly abrasive. Note that this tooth shows some deformation in its mid part. This specimen was collected associated with giant Miocene shark teeth by diving in the Cooper River, South Carolina. (Value range E).

Mammoth tusk fragments: Weathered, chalky mammoth remains are sometimes found in loess, particularly where it is thick as it is along the major rivers in the U.S. Midwest. These remains can be conspicuous because their white interiors stand out against the tan, homogeneous loess. Generally mammoth remains found in loess come from the Colombian mammoth and represent situations where these large probosidians were overwhelmed in fierce dust storms that took place after the melting of the glaciers. This wind blown dust (or silt) came from the huge amount of rock flour generated from movement of the continental glaciers. This dust or rock flour was blown onto uplands surrounding the rock flour choked rivers to form loess deposits. (Value range F, single fragment)

Mammuthus colombi: These mammoth teeth came from loess deposits outcropping along the Mississippi and Missouri rivers. These chalky and generally poorly preserved teeth are sometimes spotted and collected by construction workers where they stand out conspicuously against the tan, homogeneous loess where the bones of a mammoth, possibly one overwhelmed by a dust storm, are exposed when digging into the loess with heavy equipment. (Value range E for group).

Mammuthus sp.: A large number of Pleistocene fossil teeth such as these have come from operations trawling for bottom living fish in the North Sea adjacent to the Netherlands. These teeth are acquired by dealers and collectors in the fish markets of Amsterdam. These fossils represent the tough parts (primarily teeth) of animals that lived on an extensive continental shelf which is now under water but which was land during the periods of glaciation when sea level was considerably lower than today. Land was exposed on the continental shelves when the large amount of water required to form continental glaciers lowered sea level many hundreds of feet. (Value range F, single tooth).

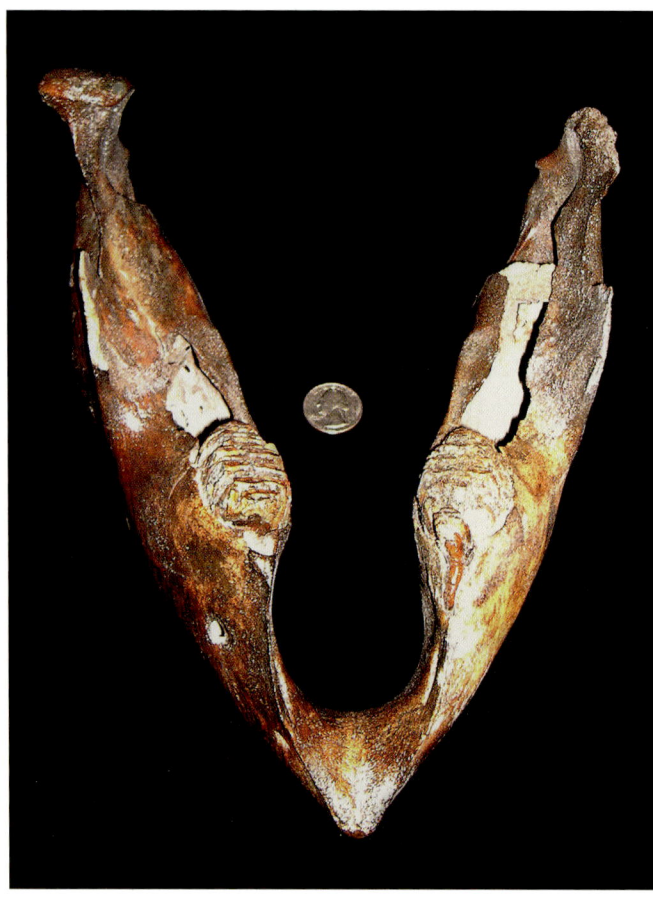

Lower jaw of a newborn, baby mammoth: A large number of well preserved mammoths (sometimes with flesh still preserved), as well as other ice age animal remains, are found in frozen ground of eastern Russia (Siberia), some of which have entered the fossil market in the last few decades. (Value range B).

Mammuthus cf. *jeffersoni*: This banana-shaped tusk came from sediments in the bed of the Grand River in northwest Missouri. Prior to the continental glaciers "pushing" the ancestral Missouri River southward to its present location at Kansas City, the ancestral Missouri cut its valley in northwest Missouri which, when the glacial ice melted, became the course of the Grand River, which now empties into the Missouri River 150 miles downstream from Kansas City. Pleistocene sediments of these river systems can locally yield a variety of ice age mammal fossils. (Value range C).

Mammuthus primigenius: A (composite) mammoth skull found near Fairbanks, Alaska, during the mining of gold-bearing, placer gravels. The large hole in the middle is for the animals trunk. This hole is not an eye socket, but has been misinterpreted as such, leading to myths of one-eyed prehistoric giants. (Value range A).

The other side of this large tusk from the Grand River of Missouri.

Strands of mammoth hair presented in a format that has been widely distributed through the fossil market. (Value range F).

Necklace made of mammoth bone beads. Collectors accumulate numerous bone fragments of mammoths and mastodons over time. One use of such fragments is to produce these unique necklaces, which occasionally show up at fossil, artifact, and mineral shows. (Value range E).

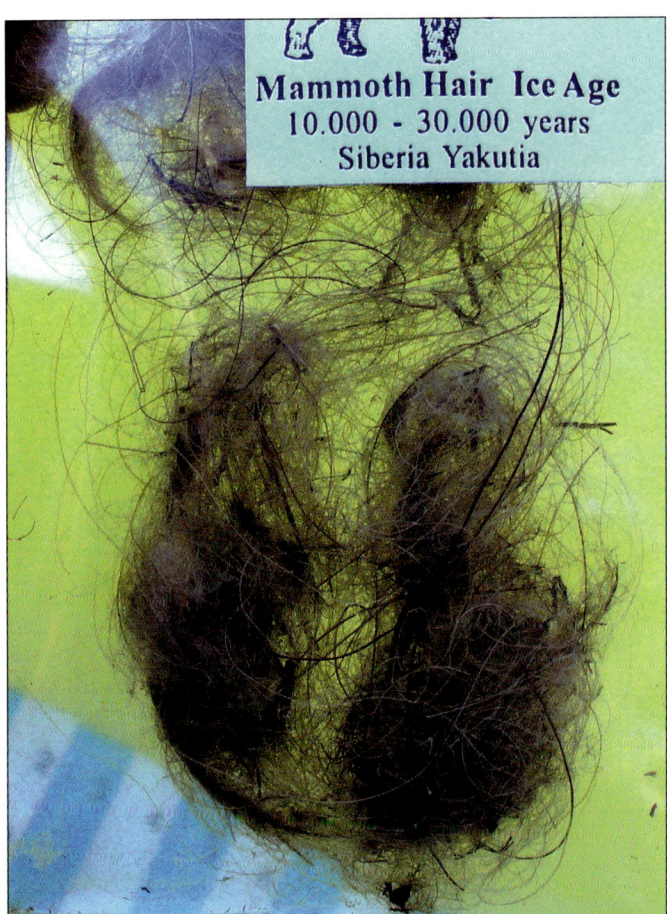

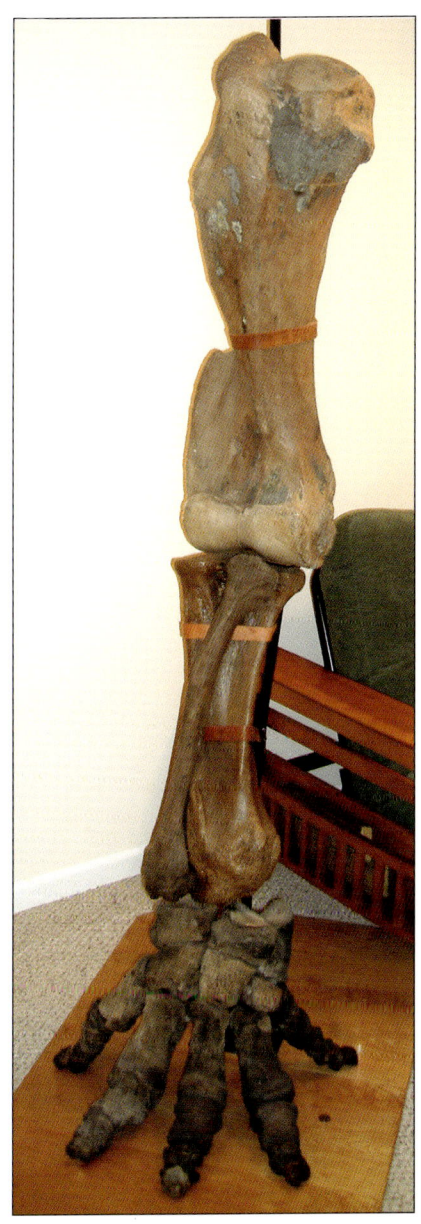

Mammoth foot: The right front foot of *Mammuthus* cf. *primigenius* reconstructed from remains found in gravel digging operations on the Kansas (Kaw) River of eastern Kansas. (Value range B).

Mammoth hair: Permafrost (frozen soil) of both Siberia and Alaska yield not only the bones of ice age animals but (at times) preserve organic tissues where such material has been preserved by deep freezing in the permafrost. Mammoth hair can be reddish in color as well as grey, as are some of the strands shown here. (Value range F).

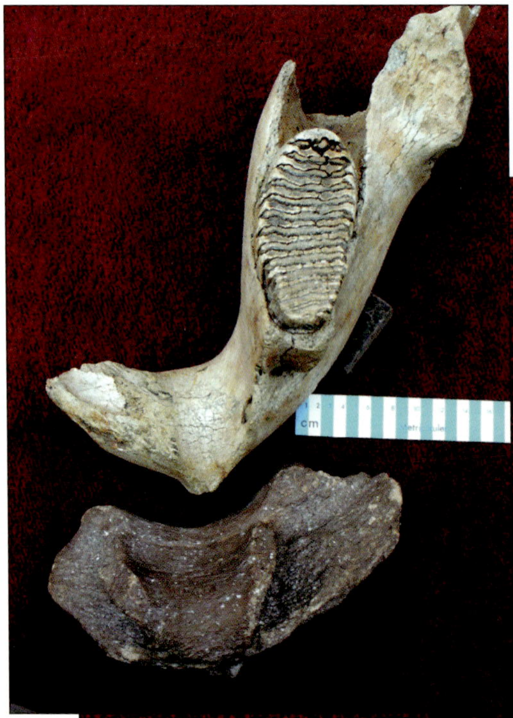

Mammoth jaws: These are partial jaws of a small, juvenile mammoth from Pleistocene sediments found in a gravel operation on the Kansas River near Bonner Springs, Kansas. (Value range D).

Close-up of mammoth jaw shown in previous photo. (Value range C).

Mammuthus primigenius: A well-preserved mammoth tooth—the grinding surface is at the top. Large mammoth and mastodon teeth, like nickel-iron meteorites, are sometimes spotted by persons who otherwise would ignore most other "rocks." Such things are spotted by persons who have no knowledge of what they have found but who notice and collect such "rocks" because they are conspicuous and odd. This specimen was spotted and collected by a farmer while fishing in eastern Arkansas. (Value range C).

137

Another large and well-preserved mammoth tooth found and collected by a local who collected it because it was a conspicuous odd "rock." (Value range C).

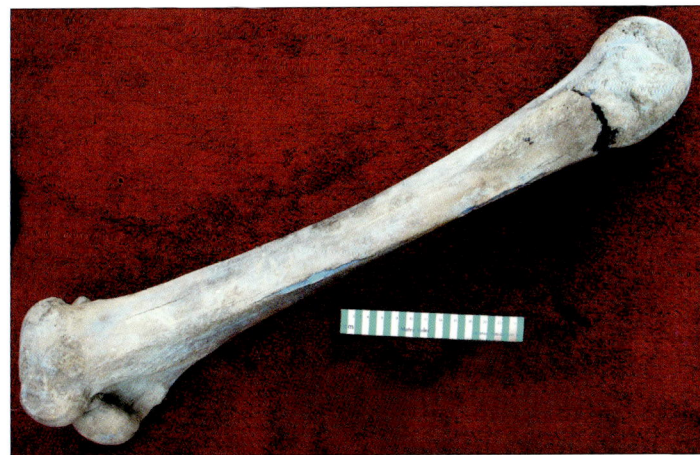

Mammoth femur: This well-preserved femur was found associated with placer gold mining activity in Alaska. Note the blue coloration, particularly in the middle part of the bone. The blue color is from the mineral vivianite, a ferrous phosphate often associated with bone bearing Pleistocene sediments.

Specimens like these abraded mammoth teeth are found as cobbles either in the beds of streams, in gravel excavation sites or on gravel bars. A search image of mammoth (and mastodon teeth) when looking on gravel bars or in streams can sometimes produce water worn teeth like these. (Value range F, single specimen).

Baby mammoth: This small **baby** (**not pigmy**) mammoth skull was found in frozen ground and, when preserved in this manner, resembles modern bone. There is absolutely no mineralization or mineral stain on it, which is usually the case with fossil bone. (Value range C).

A similar group of worn mammoth teeth found as stream cobbles on gravel bars. Mississippi River gravel bars, New Madrid County, Missouri. (Value range E for group).

Tooth of a small (juvenile) mammoth: This tooth came from a gravel bar on the Grand River of northwestern Missouri. (Value range F).

deserts of Africa, where the plants are extremely nourishing, produce the largest and fiercest animals; and, perhaps, for a contrary reason, America is found not to produce such large animals as are seen in the ancient continent. But, whatever be the reason, the fact is certain, that while America exceeds us in the size of its reptiles of all kinds, it is far inferior in its quadruped productions. Thus, for instance, the largest animal of that country is the tapir, which can by no means be compared to the elephant of Africa. Its beasts of prey, also, are divested of that strength and courage which is so dangerous in this part of the world. The American lion, tyger, and leopard, if such diminutive creatures deserve these names, are neither so fierce nor so valiant as those of Africa and Asia. The tyger of Bengal has been seen to measure twelve feet in length, without including the tail; whereas the American tyger seldom exceeds three. This difference obtains still more in the other animals of that country, so that some have been of opinion * that all

1795 excerpt on the lack of large mammals in modern North America. An interesting side issue regarding mammoths was the knowledge of their bones and teeth as early as the mid-eighteenth century at Big Bone Lick in the territory of Kentucky. The origin of these large fossil bones and teeth remained totally puzzling to all concerned, which included Benjamin Franklin, who once visited the site. Big Bone Lick was visited later by Thomas Jefferson, who identified the remains as being from elephants after one of his slaves, familiar with elephants in his native Africa, identified the teeth as coming from such. The text shown here was originally written by the French naturalist Comte de Buffon in the mid-1700s and paraphrased in 1795 by Oliver Goldsmith. This text was familiar to both Franklin and Jefferson, the latter who questioned its statement that "America is found not to produce such large animals as are seen in the ancient continent(s)" (viz. Asia and Africa). Jefferson, attempting to prove his belief that large animals like these did indeed still exist in North America, instructed William Clark and M. Lewis to be on the lookout for living, hairy elephants on their famous expedition at the beginning of the nineteenth century.

Juvenile mammoth placed against a baby mammoth outline. Kaw River, eastern Kansas. Specimen was on display at MAPS Expo 2009.

Mastodons

The mastodon is probably the second best known of the Pleistocene megafauna. Mastodons are unique to the Pleistocene of North America; they were more woodland browsers then were mammoths, which lived in the more open areas. Mastodons were especially characteristic of the late Pleistocene of North America.

Bronze model of a mammoth from a "dinosaur" set sold in the 1950s at natural history museums. With such models the public is often unaware of the vast time span which separates the Pleistocene megafauna from the Mesozoic Era. Sometimes the word dinosaur is (incorrectly) used to represent any extinct, large vertebrate animal. (Value range F).

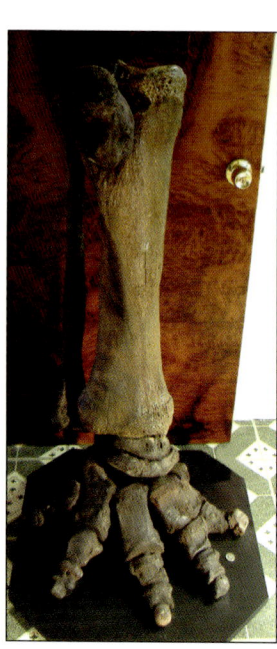

Mastodon foot: The bones of this massive (reconstructed) mastodon foot came from gravel excavations along the Kansas (Kaw) River near Bonner Springs, Kansas. (Value range B).

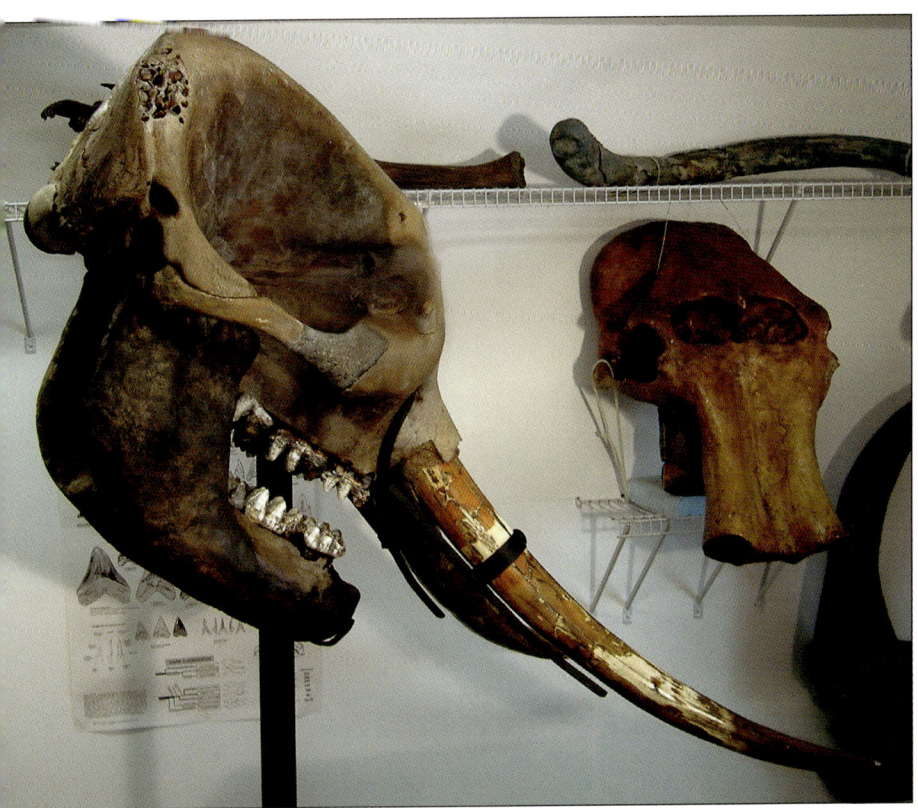

Mastodon americanus: A mastodon skull found in a peat bog of central Michigan. Draining or excavation in the organic muck of peat bogs sometimes can yield the bones of large elephants—particularly mastodons. When the bones of these large animals are found in excavations, they sometimes produce a local newsworthy event. The finding of mammoth and mastodon bones is really not that rare, but such publicity is desirable as it assures that the bones being found are collected rather than being ignored and reburied—the fate of many nice fossils. *Courtesy of Rich Hagar.*

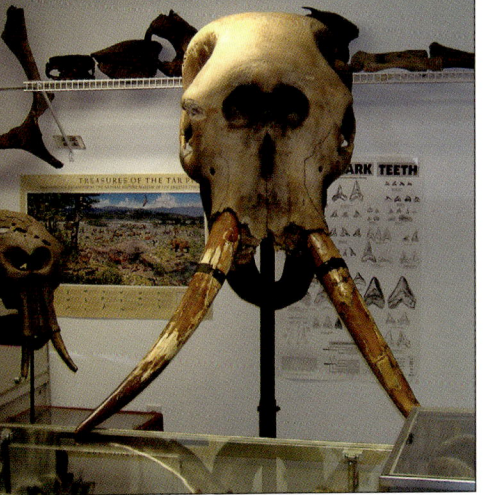

Front view of the Michigan specimen shown in the previous photo.

Richard Hagar, who acquired, prepared, and mounted this Michigan specimen, standing next to it.

Mastodon tusks: This long, slender tusk came from a pond excavation near Farmington, Illinois. Illinois has provided a large number of excellent mastodon skeletons—check out the Illinois State Museum web site. The shorter tusk fragment came from Missouri River sediments. It is partially enclosed in a large ferruginous, gravel bearing concretion that was dug up during gravel mining excavations. (Value range C).

Close-up of a portion of a mastodon tusk preserved in a ferruginous concretion. Missouri River gravels, western Missouri (Kansas City area). (Value range E).

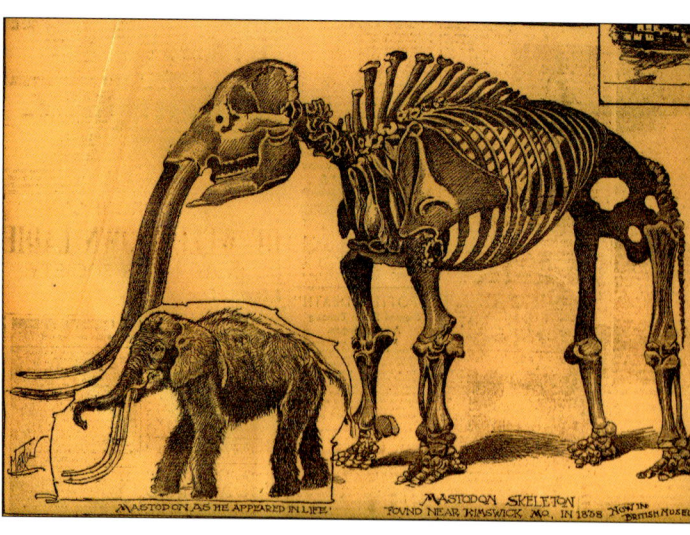

Mastodon reconstruction: Prior to the Louisiana Purchase Exposition (St. Louis World's Fair), considerably publicity was directed toward places outside of the St. Louis area that fair goers could visit on one day excursions by train. One of these was the Koch Mastodon site near Kimmswick, Missouri, now the site of Missouri's Mastodon Park. This article came from a local St. Louis newspaper from 1899.

Mastodon jaw section: A locality south of St. Louis on Rock Creek yielded numerous mastodon bones in the early nineteenth century. In the late 1830s, a German immigrant by the name of Albrecht Koch excavated a number of these bones, reconstructing them into a strange looking beast, which he called the Missouri Leviathan or the Missourium. He displayed his creation in a "dime museum" on the St. Louis riverfront where the present day arch is located. Koch's museum was disbanded with the beginning of the Civil War when the institution of martial law came to St. Louis. He then sold his Missourium to the British Museum in London where it was correctly reconstructed as a large probosidian—a mastodon. Missouri's Mastodon Historic Site is located today where Koch found many of his mastodon bones. This jaw section, with its partially erupted tooth, was collected from Koch's Rock Creek locality sometime in the late nineteenth century, possibly just before or during the time of the St. Louis (1904) World's Fair (Louisiana Purchase Exposition), where a lot of activity took place at the locality as it was a designation for fair participants to visit by daily train excursions.

Top (dorsal) view of the same mastodon jaw with Snuggles.

Albrecht Koch's mastodon locality or "fossil farm" on Rock Creek illustrated in an 1899 newspaper article that generated interest in the bone site where excursions from the 1904 World's Fair focused on it. This is now Missouri's "Mastodon State Historic Site" located on I-55 just south of the St. Louis Metropolitan area.

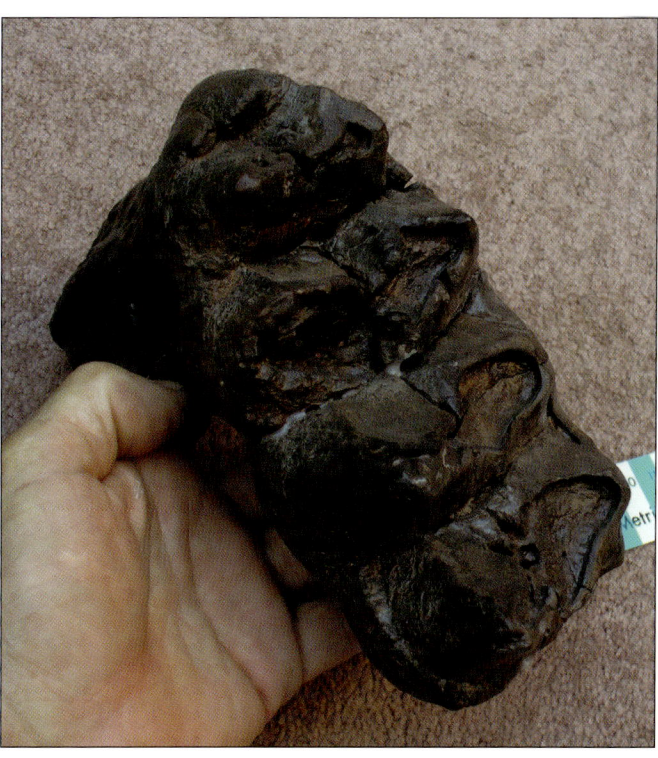

Mastodon tooth: Fossils like this one, which came from a Florida river, are generally stained and impregnated with black manganese dioxide, they may show other forms of mineralization as well. Large numbers of ice age fossils have come from Florida as large populations of these animals existed on its peninsula during the Pleistocene and their bones and teeth become concentrated in its rivers. These bones come from sediments along the rivers, from the riverbeds themselves, and also from Pleistocene sediments uncovered either during construction or in phosphate mining. From Santa Fe River, Florida. (Value range C).

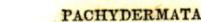

Big Bone Lick in Kentucky was one of the first discovered concentrations of mammoth and mastodon bones found in North America. It was well known by the late 1700s for its concentration of these large bones. Fossil bones from the Big Bone Lick locality were investigated by and information on them published in the scientific literature by Thomas Jefferson in the early nineteenth century. This is a paragraph discussing the site and its fossils in a 1854 text book on geology by James D. Dana, a pioneer American geologist best known today for his work in mineralogy (Dana's Mineral Numbers and Index).

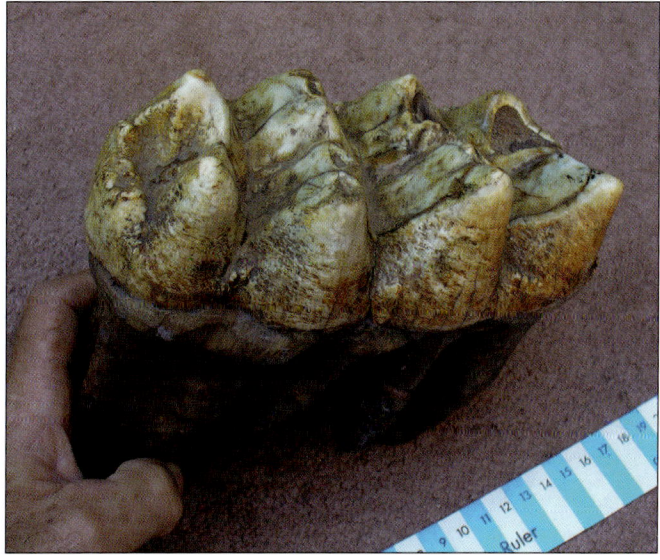

Mastodon tooth, Southeastern Missouri boot-heel: Extensive Pleistocene river deposits formed from the ancestral Mississippi River occur in Missouri's boot-heel region of the southeastern portion of the state. These sediments sometimes can yield nice vertebrate fossils. (Value range C).

Mastodon-lower palate: This came from a large Mastodon, the bones of which were found in gravel excavations on the Kansas (Kaw) River of eastern Kansas. (Value range B).

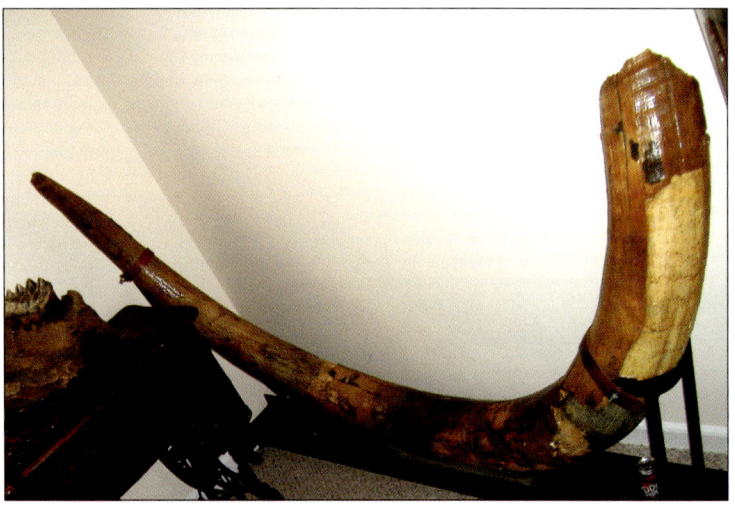

This tusk came from a very large mastodon found associated with the above palate in a gravel excavation on the Kansas River in eastern Kansas. Much of the flood plain of the Missouri and Mississippi rivers, as well as their major tributaries, are composed of late Pleistocene sediments in which are found the bones, teeth, and tusks of mastodons and mammoths. (Value range A).

Mastodon bone fragments with vivianite stain: Vivianite (a ferrous phosphate mineral) can coat Pleistocene bones, sometimes turning them blue. The intense blue of this mineral can resemble the mineral azurite, which is a copper carbonate. Pleistocene swamp deposits, eastern Missouri. (Value range G for group).

Ball at the end of a mastodon femur: These fit into a "ball and socket" at the femur-tibia junction and resemble a "fossil bowling ball." Pleistocene swamp deposits, Coldwater Creek, eastern Missouri. (Value range F).

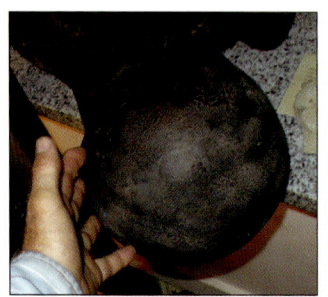

Another mastodon "bowling ball."

Carnivores

Carnivores generally are rarer vertebrate fossils than are herbivores as, being more intelligent, they often avoided situations which might otherwise trap less intelligent animals and preserve their bones as fossils. Bones of cave bears in Europe are an exception to this rule, occurring in such abundance in many caves of eastern Europe that in the nineteenth century their bones were mined from some of these cave deposits, ground up, and used as fertilizer.

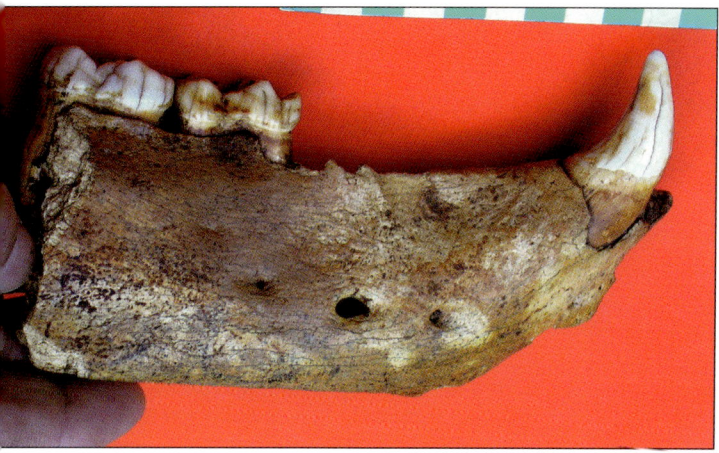

Arctodus pristinus: Short faced bear. This is the lower jaw of an extinct bear that was widespread in North America during the Pleistocene Epoch. It was larger than the grizzly bear (Value range E as a well made cast).

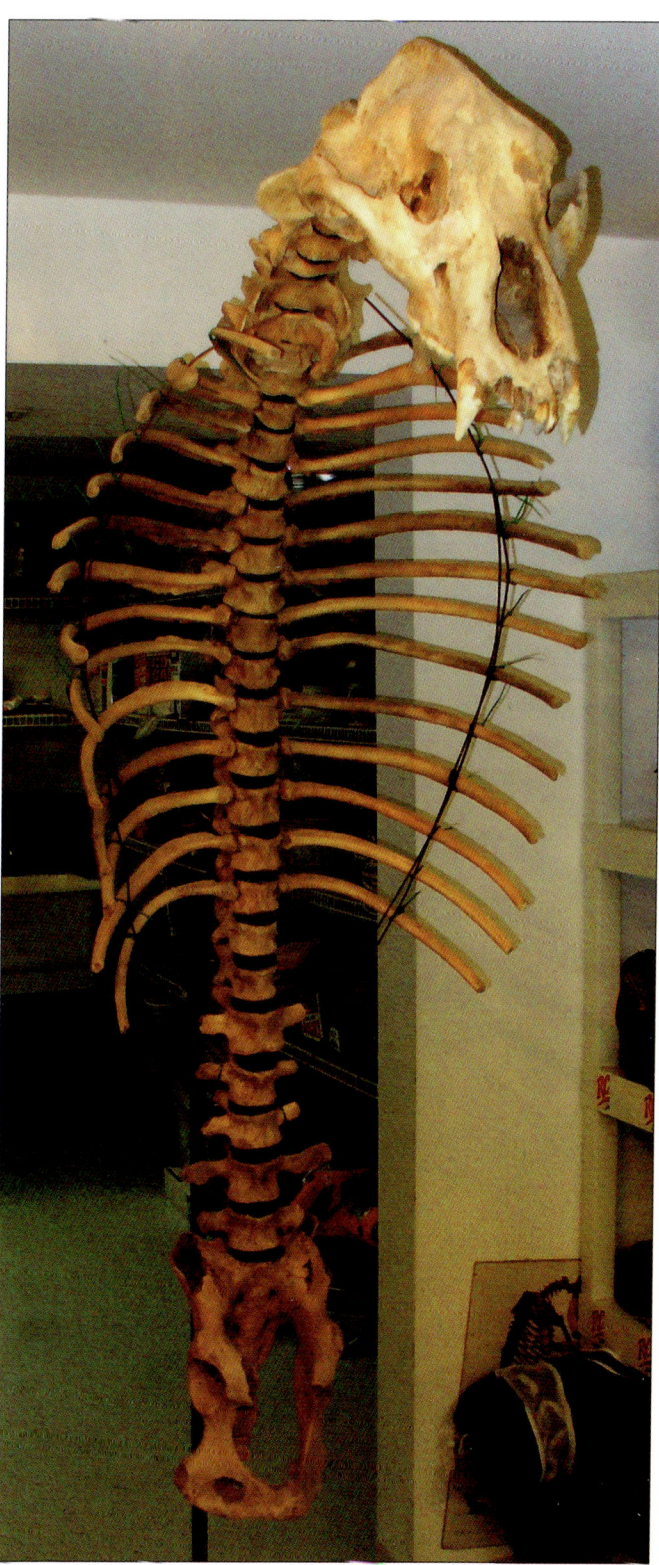

Ursus spelaeus: Cave bear, partial mounted skeleton. A relatively large number of remains of the European cave bear have come onto the fossil market through the Tucson Arizona show, including this one. Many of these cave bear fossils were collected from caves in the Carpathian Mountains of Romania. (Value range B).

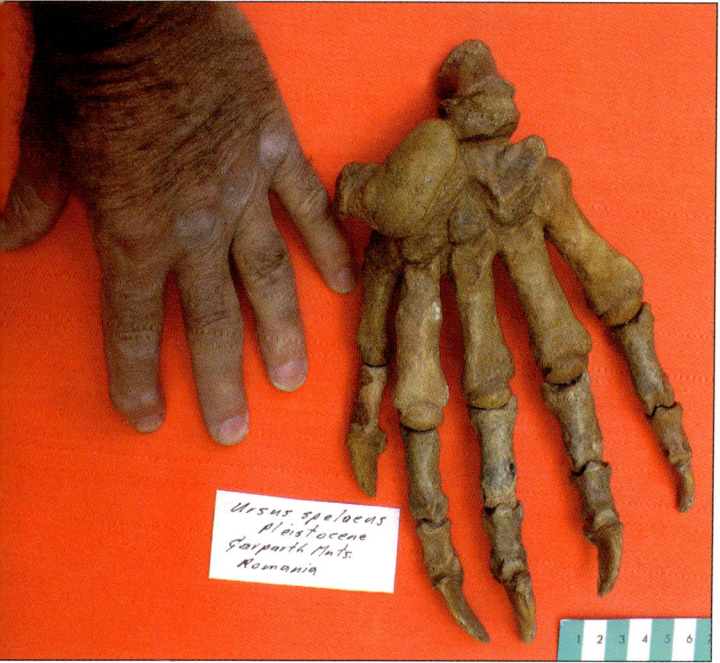

Ursus spelaeus: A reconstructed paw of a cave bear that came from cave deposits of the Carpathian Mountains, Romania. (Value range E).

Ursus spelaeus: Cave bear partial lower jaw from Romania. (Value range F).

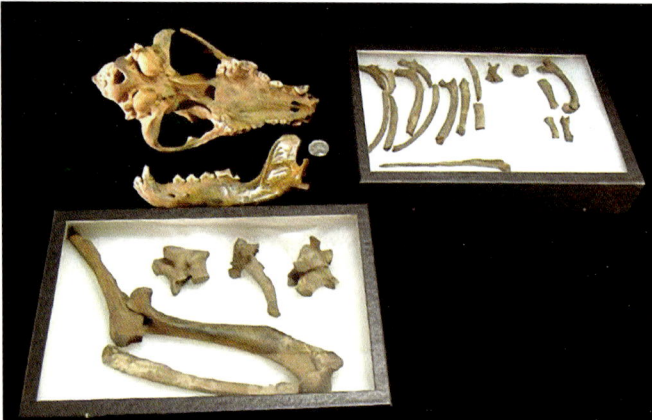

Canis dirus (dire wolf): One of the large Pleistocene carnivores. This specimen came from gravels of an Alaskan placer deposit and was found associated with mammoth bones.

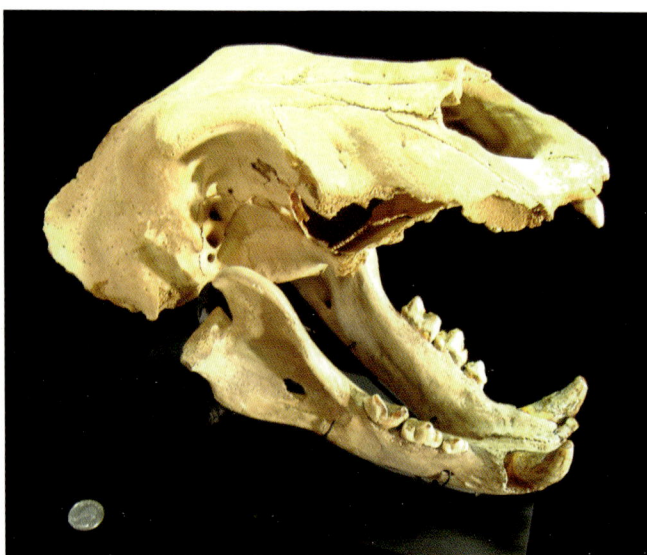

American lion: A large cat that came from gold bearing placer deposits of Alaska. Absolutely no mineralization occurs with the original bone—a characteristic of Pleistocene bone found in frozen ground. Under ideal conditions some of the flesh and hair of these animals can even be preserved, sometimes as a sort of mummy. Mining of gold bearing gravels in Alaska (placer gold deposits) has produced a large number of well preserved ice age animals.

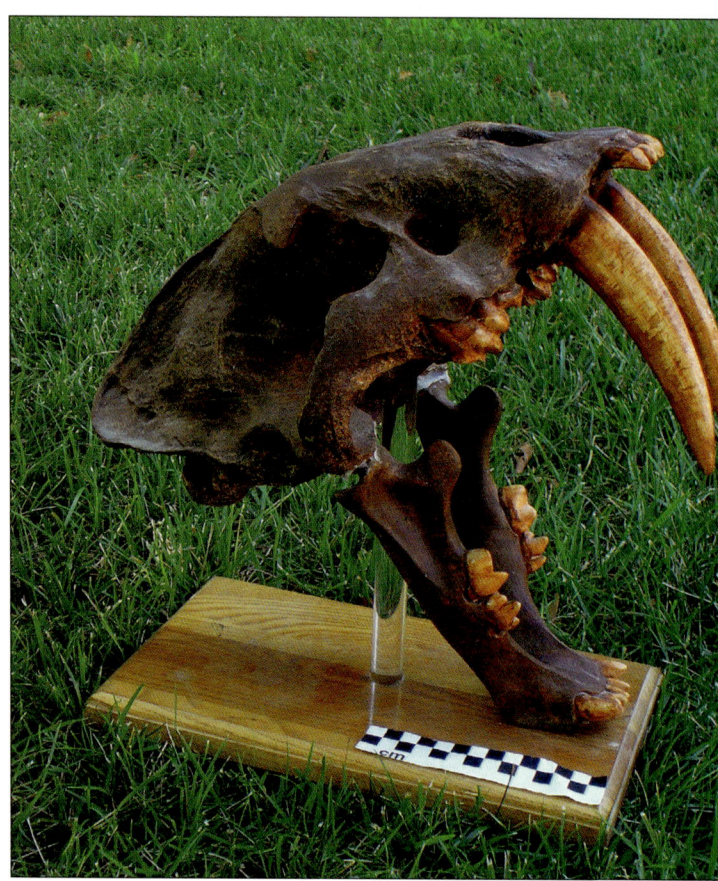

Smilodon sp.: The saber-toothed cat is one of the most desirable fossils of the Pleistocene Epoch. This is a cast of one of the many specimens of this extinct carnivore, which come from the La Brea tar pits of Los Angeles, California. (Value range E, as cast).

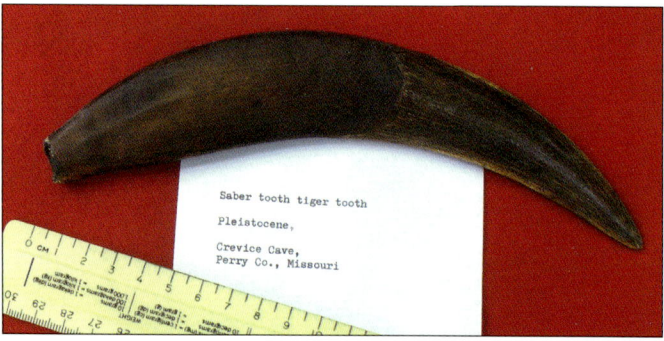

Large canine tooth of a saber-toothed cat. Crevice Cave, Perry Co., Mo. (Value range C).

Worn teeth of Pleistocene carnivores: Teeth are some of the most durable parts of vertebrate animals. These abraded teeth of fossil carnivores were found on gravel bars of the Missouri River. (Value range F).

Coprolite: A phosphate replaced "turd," probably from a carnivore. Late Pleistocene, swamp deposits, Coldwater Creek, eastern Missouri. (Value range G).

Tracks and Trackways

Caves can preserve **tracks** and **trackways** on their floors as well as the bones of Pleistocene vertebrates.

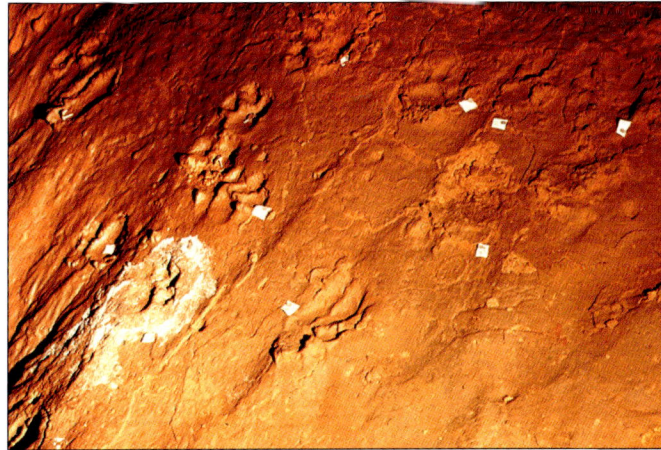

Large felid (cat) tracks in the same cave as the small raccoon tracks: A jaguar or the American Lion may have made these relatively large cat tracks. They are about the same age as are the small raccoon tracks—that is some 30,000 years old. Caves can preserve ancient trackways on the surface of their underground streams, which deposited and flowed over red clay. Later, the stream cuts downward into bedrock but most of the original clay floor, containing the tracks, stays high and dry, thus preserving the tracks. Virgin caves in particular should be watched for these tracks and care taken not to trample any smooth clay surfaces when traversing through such a cave. Well-traveled caves generally did have such tracks at one time but the trampling of visitors usually has destroyed these. The white material on the track at the left is plaster residue left behind when a plaster mold was made of this track.

The floors of caves, especially caves having no (current) natural entrance, sometimes preserve tracks and trackways of Pleistocene (and Holocene) animals. If such a cave is found, care should be taken not to trample any smooth clay surfaces—they might contain such trace fossils. Here the author is in a much-trampled cave along a popular Ozark float stream. Any tracks that might have existed here have long been trampled and destroyed. Note the mud splatter on the large stalagmite, presumably created by stomping cave visitors.

Casts of the plaster mold taken from the cat track (the one with the white plaster residue). The original tracks in this cave were severely damaged recently by greatly increased amounts of water entering the cave as a consequence of recent surface development (a subdivision) and its diversion of runoff into cave-connected sinkholes. Photos and replicas of the original tracks now appear to be all that remains of them from this particular cave. The cave now also contains sewer gas generated from septic systems associated with the surface development. Brome Moore Cave, Perry Co., Missouri.

Fossil raccoon trackways on the floor of an Ozark cave: Water once flowed over this clay surface which, when it was wet and soft, received the tracks of visiting animals coming into the cave through now blocked entrances. Since these tracks were made, the stream that flowed through this passageway has cut some 25 feet downward into the bedrock (dolomite) of the cave. At a downward cutting rate of .01 inch/year by the stream, it is estimated that to cut twenty-five feet (300 inches) would take at least 30,000 years. (.01 in/yr x 300 in = 30,000 yrs). This is a **minimum** age for these raccoon tracks, as well as for the cat tracks in the following photos, which occur in the same cave.

Another angle of the cat-track passage in Brome Moore Cave, Perry Co., Missouri. (Photo taken in the winter, 1970).

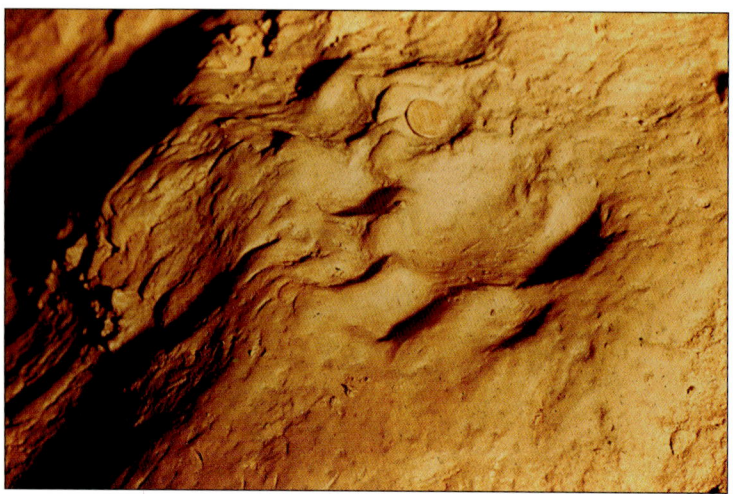

And another, but different view!

Close-up of a cat track that has been partially covered (at the left) by a layer of dripstone. Brome Moore Cave, Perry Co., Missouri.

Stone Tools and Humans

Some of the rarest of Pleistocene fossils are the remains of ice age humans. Humans appear to be a product of the ice age, but skeletal remains, like those of Neanderthal man, are rare and coveted fossils. Stone tools made by Pleistocene man are more common, but usually they are difficult to date accurately.

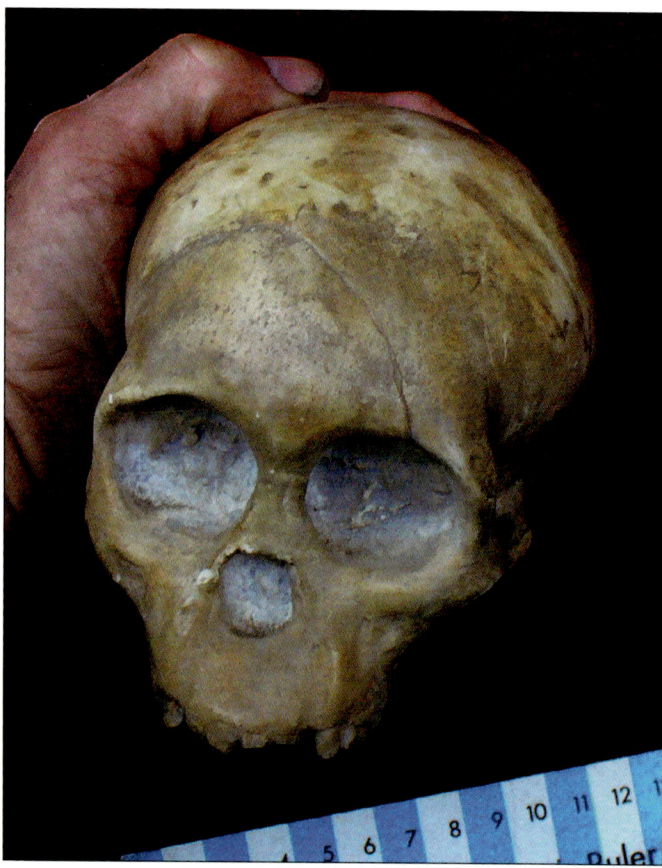

Australopithecus africanus (cast, Taung child): This mid Pleistocene hominid was found embedded in dripstone in a South African cave in the 1920s. It was one of the first indications suggesting that the earliest fossil record of man might be found in Africa rather than in Asia—the current hypothesis at the time. Fossil hominids are one of the rarest and most coveted of fossils. The genus *Australopithecus* was similar in many ways to a chimpanzee. The next evolutionary stage leading toward modern man was *Homo habilis* (the genus Homo now being used to indicate more human-like features). Somewhere between *Australopithecus* and the next stages in human evolution, *Homo erectus* and *Homo sapiens*, intelligence evolved. (Value range E for a fine cast).

A large Acheulean "hand ax," Northwest Africa: Hand axes like this are found with some frequency scattered throughout the dry deserts of central Africa and are also found in China, India, southeast Asia, and in Europe. They are called hand axes, although it is not known in just what manner they were used. Such tools are considered fossils as they appear to have been made some 500,000 years ago. They have entered the fossil market in quantity, coming through Moroccan fossil dealers who acquire them from locals who collect them from the desert of northwest Africa. Such hand axes are usually composed of quartzite and many exhibit the effects of sandblasting while lying for centuries in the desert sands. Like many artifacts acquired from dealers, however, there is a degree of uncertainty as to the authenticity of some of them. Objects like this can be made and then "aged" fraudulently. This specimen came through a reputable dealer at the Tucson, Arizona, show. It is a particularly large "hand ax." (Value range C).

Bottom: Olduvai Gorge hand ax; Top: "hand ax" from the Sahara Desert, acquired at the Tucson, Arizona, fossil-mineral-gem show. Inclusion of these tools in this chapter is done as these stone tools are older than 10,000 years, the "cut off" point for geologic time and also the "cut off point" as to what is or is not a fossil (greater then 10,000 years it's a fossil, if younger, it is **not**!) Stone tools made by early humans are much more common than are related skeletal remains; these triangular "hand axes" are among the most accessible and available of Pleistocene hominid fossils.

Hand ax: A documented tool made by "hand ax" man from Olduvai Gorge, Tanzania, Africa. The specimen shows an iron oxide patina as well as iron oxide "buttons," which formed in a dry desert environment over time. (The "blueberries" found associated with Martian rocks had a similar origin). Olduvai Gorge became a significant hominid site through the work of Louis and Mary Leakey in the 1960s. This specimen, along with others, was acquired sometime in the 1950s by Geological Enterprises—possibly through the Leakey's. The bones and tools of "Hand ax man" occur in the upper-most layers of Olduvai. The lowest layers of the gorge contain very primitive tools of Australopithecines, which date back to the earliest Pleistocene or to the latest Pliocene. (Value range B for documentation and "pedigree").

A wood cut from a late nineteenth century book on the ice age showing a possible Pleistocene(?) "hand ax" found in eastern North America. This rendering is compared to a hand ax from the Sahara Desert of northwest Africa. In N.W. Africa such hand axes have been documented to be 400,000+ years old and hence they are well into the ice age and also are bona fide fossils. The earliest appearance of humans in North America is still a live question. An interesting (1999) article on this "sticky" issue is in *Newsweek*, April 1999, entitled "Who were the first American's?"

A wood cut sketch of a "quartz implement" found in eastern North America and illustrated in *Man and the Glacial Period* by G. Frederick Wright, 1896. To the left is a "worked flint" found in northern Florida of a similar type as illustrated in the wood cut.

Stone tools (flint paleoliths) collected from rivers in eastern North America. Such tools, if found in place in undoubted Pleistocene sediments, would help in documenting the very "sticky" issue as to the time of appearance of humans in North America. A considerable amount of confusion still surrounds this particular issue. Dating of these relatively common stone artifacts is, however, exceedingly difficult. (Value range F, single flat tool).

Stone tools ("paleoliths") from eastern North America compared with similar tools reported found in glacial sediments. Wood cut illustration from Wright, 1896, *Man and the Glacial Period*, D. Appleton and Co.

Paleolithic picks or scrapers, southern England: These stone tools come from gravels presumed to be late Pleistocene in age. Gravels of the same age containing similar tools were the source of the famous Piltdown Man forgery of the early twentieth century (see S. J. Gould, *The Piltdown Conspiracy*). The bones of the "Piltdown Man" were fraudulently "salted" in the gravels. The remains were a combination of human and gorilla bones, which were then "found" a second time, perpetrating the fraud that became the Piltdown Man forgery. Note the ferric oxide stains on the top specimen—generally this is a good indication that the flint tools are really authentic and ancient as it takes centuries (usually) to acquire such a patina. Modern day forgers have found ways to reproduce this, however. Even with stone tools and artifacts and a guarantee of their authenticity, it's still caveat emptor. (Value range C).

Clovis Point: One of the earliest of stone tools found in North America are Clovis Points. These belong the Archaic (cultural) stage as identified by North American archeology, a period slightly less than 12,000 years ago. The Clovis hunters were hunters of mammoths. Clovis points are often found with the bones of the mammoth. Clovis points also are true fossil artifacts, being older than 10,000 years, the cut-off point that is generally given for the end of the Pleistocene Epoch and of geologic time. Both pre Clovis artifacts and skeletal materials are known, which "relate" to the contentious issue of who were the first humans in North America. (Value range B).

Bibliography

Begley, Sharon and Andrew Murr, 1999. "The New Scientific War over a 10,000 Year Old Puzzle. Who Were the First Americans." *Newsweek*, April 26, 1999.

Gould, Stephen J., 1977. "The Misnamed, Mistreated and Misunderstood Irish Elk" *in Ever Since Darwin*. W. W Norton and Co., New-London. ISBN 0-39300917-3.

_____. 1980. *The Piltdown Conspiracy in Hen's Teeth and Horse's Toes, Further Reflections in Natural History*. W. W. Norton and Co., New York-London. ISBN 0-393-30200-8.

_____.1982. "Piltdown Revisited" *in The Panda's Thumb—More Reflections in Natural History*. W. W. Norton and Co. New York-London. ISBN 0-393-30023-4.

Hedeen, Stanley, 2008. *Big Bone Lick, the Cradle of American Paleontology*. The University Press of Kentucky.

Lange, Ian M., 2002. Ice age *Mammals of North America; A Guide to the Big, the Hairy and the Bizarre*. Mountain Press Publishing Co. Missoula Montana. ISBN-87842-403-2.

Lister, Adrian and Paul Bahn. 1994. *Mammoths*. Macmillan Publishers, New York. ISBN 0-02-572983-3.

Mayor, Adrienne, 2000. *The First Fossil Hunters; Paleontology in Greek and Roman Times*. Princeton University Press, ISBN-691-08977-9.

McMillan, R. Bruce, 2010. "The Discovery of Fossil Vertebrates on Missouri's Western Frontier." earth *Sciences History*, Vol. 29, No. 1. *Journal of the History of earth Sciences Society*.

Mehl, Maurice G., 1962. *Missouri's ice age Mammals*. Educational Series No. 1, Geological Survey of Missouri-Rolla.

Weaver, Kenneth F., 1985. "Stones, Bones and Early Man; The Search for Our Ancestors." *National Geographic* Vol. 168, No. 5, November 1985.

Chapter Eleven
The Holocene or Recent Epoch

This is the youngest portion of the geological time scale and as it encompasses recorded history technically, it really **isn't a part of geologic time at all**. The Holocene is the time from the **end** of the ice age, some 10,000 years ago, to the **present**. It was during this time period that civilization developed. The Holocene represents the time of this history! Antiquities and artifacts from the Holocene are not fossils; they are **not** old enough. (Bona fide fossils have to be at least 10,000 years old). In the Holocene are found various objects that some persons confuse with fossils—these being artifacts like arrowheads, scrapers, and pottery. Later in the Holocene come historical items of a vast variety, some of which, *many* of which, if they are old enough are collectable.

It might be reiterated again that items illustrated in this chapter are not fossils, but rather most fall under the field of archeology. In many instances (at least in the U.S.), considerable confusion sometimes exists as to what constitutes paleontology (with its fossils) and what constitutes archeology. First of all, **archeology deals with the antiquity of humanity and humanity alone**. Archeologists do not work (professionally at least) with dinosaurs or for that matter with any other organisms unless they are in a context related to humans. Most archeology is also concerned with the Holocene, the time period from 10,000 years ago to the present. That is not to say that archeologists don't work with older records—the human record goes well into the Pleistocene and Pleistocene archeology is just as valid as Pleistocene Paleontology. It is usually in the Pleistocene that paleontology and archeology meet and potentially converge. Archeology has some validity even with pre-Pleistocene hominids and other higher primates, the problem being here in defining exactly what is and what is not hominid (human) related—and where in the fossil record "man" begins. The important distinction again being that archeology deals exclusively with humans. With this in mind, the author recalls how often archeology and paleontology are confused in the news media—particularly with post-*Jurassic Park* news coverage. It is sometimes commonplace to see or hear statements like, "The archeologists are working with this dinosaur" or that, "As a person interested in fossils, I bet you'd love to go to Egypt and work with mummy's." This appears particularly prevalent with religious persons who seem unable to understand the concept of megatime—or perhaps don't want to understand it. The author has periodically "locked horns" with some Young earth Creationists on this matter, that especially vocal and dogmatic group of Christian fundamentalists who believe in a **literal six days of Creation**.

Worked flints, Santa Fe River, Florida: These primitive stone tools were made some 2,000 years ago. They are **not** old enough to be bona fide fossils. In many ways they resemble worked flints found in early or mid-Pleistocene age strata and made by extinct species of humans, viz. *Homo erectus*. Primitiveness in stone tool making represents a **cultural** or **technological** stage in human development and as this varies in time different cultural stages appear at different times in different groups. Large numbers of these primitive worked flints from North America have entered the collector's market where they have been collected from rivers in Florida and other parts of the eastern U.S. by diver-collectors. Some of these "primitive" stone tools may in fact be rock cores or "blanks" intended to be used in making tools. Distinguishing a real primitive stone tool from a rock core or blank without the object being found in place in its original sediment, is quite difficult if not impossible. (Value range G, single specimen).

Worked flints, eastern North America: Crude flint tools, probably used as scrapers were made by North American Indians. Such artifacts in the nineteenth century could locally litter the ground at some places; however, today, although less common, they can still be found where they form a link between us and the inhabitants of North America 600 to 3,000 years ago. (Value range F for group).

A group of arrowheads (projectile points) and scrapers made from 800 to 3,000 years ago by Native Americans of the Midwest U.S. Such well made stone tools contrast with the crudely made stone tools shown previously. **A**. Lancoelate point, Late Archaic, Salt River, NE Missouri. **B**. Scraper, Woodland Period, Lincoln Co., Missouri. **C**. Drill—made from an arrowhead, Callaway Co., Missouri. **D**. Bird point, Ferguson, St. Louis Co., Missouri. **E**. Etley point, St. Louis Co., Missouri. **F**. Dovetail point (St. Charles Dovetail), Early Archaic, Pike Co., Missouri. **G**. Woodland common point, St. Charles, Missouri. **H**. Woodland, Missouri River. **I**. Etley point, Late Archaic, Lincoln Co., Missouri. **J**. Etley point, Late Archaic, Lincoln Co., Missouri. **K**. Snyder type point, Crawford Co., Missouri. **L**. Snyder point, St. Charles Co., Missouri. **M**. Etley point, St. Charles Co., Missouri. **N**. Knife, Callaway Co., Missouri. (Value range C for group).

Worked flints, Northwest Africa (Mauritania): These stone tools, collected from the deserts of east-central Africa, are typical of stone tools found worldwide. They probably are five to six thousand years old, but they exhibit an amount of craftsmanship equal to or greater than the younger tools shown in the previous photos. This is a consequence of the group in Africa that produced these tools having reached a higher cultural level at an earlier time than did the makers of the North American tools. (Value range C for group).

Close-up of a single stone spear head from Mauritania. (Value range C for group).

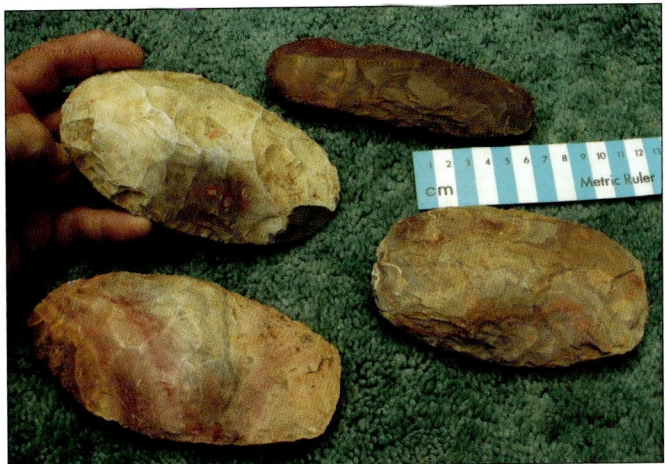

Northwest African hand tools (adz): There is a similarity in stone tools worldwide. Stone tool making reached a higher level of development in the "Old World" earlier than in the "New," and these are a few thousands of years older then are comparable tools found in America by Native Americans (Indians). (Value range D for group).

Celt: This pre-form was made by a member of the Mississippian Native American culture, some 1,000 years ago. It is made from a glacial cobble composed of hard, crystalline igneous rock. The igneous cobbles used by Native Americans to make these were brought down from Canada or northern Minnesota by continental glaciers during the Pleistocene Epoch. (Value range E).

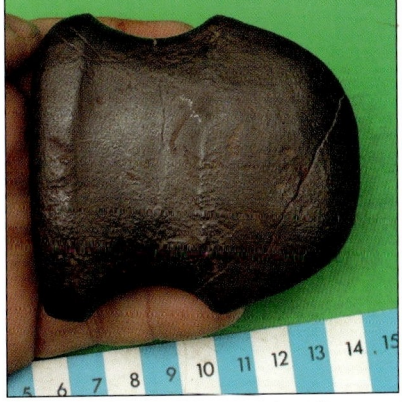

Hematite ax head: Hematite (ferric oxide) was a preferred material for Native Americans with which to make ax heads until the eighteenth century when iron tools (including iron ax heads) became available through trade with settlers. Hematite ax heads usually were made from cobbles of this material derived from Paleozoic rocks of the Midwest U.S. Such cobbles can often be found in the beds of small streams. In the area where this originated, Lewis County, Missouri, masses of hematite from which such ax heads were made come from shale beds of Middle Pennsylvanian age and are found in creeks. There is question however about the authenticity of this and many other hematite ax heads which show up on the rock and artifact market. As with many artifacts, as well as with some fossils, one has to be sure of what one is doing—it's **Caveat Emptor**.

Arrowheads and scrapers: Probably the best-known stone artifacts from the Holocene are arrowheads and scrapers. These came from eastern Missouri, the black one made from a distinctive black flint (Pitkin chert) from the southern Ozark Mountains of northern Arkansas. (Value range E for group).

Arrowheads and scrapers: A group of stone tools between 800 and 1,200 years old from eastern Missouri found by the author as a child. These came from what is now the St. Louis metropolitan area, but which 800 years ago was an area peripheral to a sizeable Native American community known as Cahokia Mound, which was located on the flood plain of the Mississippi River in what is now western Illinois. Because of the presence of numerous Indian mounds in the nineteenth century St. Louis region, St. Louis used to be known as mound city. (Value range F for group).

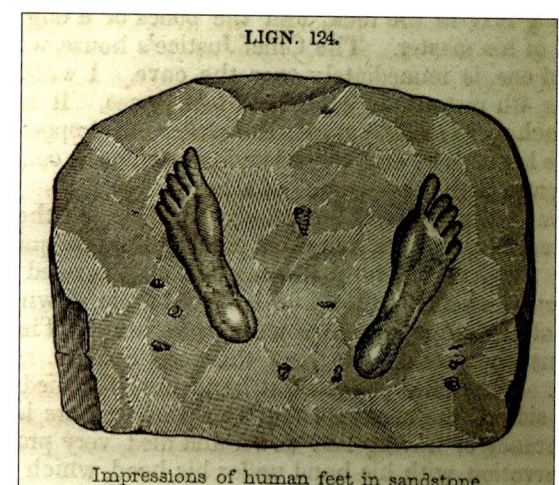
Impressions of human feet in sandstone.

In the early nineteenth century when modern geology, with its emphasis on mega-time, strata, and fossils was developing, two critical pieces of information remained unclear—the antiquity of man and the antiquity of hard (lithified) beds of sedimentary rock. As a consequence of this muddy state of understandings, considerable effort went into trying to document evidence of man occurring in beds of hard sedimentary rock. South of St. Louis, Missouri, along the Mississippi River, occur layers of hard limestone and sandstone which currently can be seen on the west side of the river where I-255 crosses the Mississippi. On one of these layers human footprints were discovered and these were cited as proof that man was in existence and walked on the soft sediment to make these tracks shortly after it was deposited. This occurrence was published in Silliman's Journal in 1821 with this woodcut (lignograph). Its author suggesting these "footprints" being proof that humans existed when these beds were deposited.

Snake and Potato-like maces: Petroglyph site in Washington State Park, De Soto, Missouri. Petroglyphs (petro = stone, glyph = writing) were carved on rock surfaces of eastern Missouri by Native Americans 800 to 1,200 years ago, probably during the peak of what is referred to by archeologists as the **Mississippian Culture** (not to be confused with the Mississippian Period of the geologic time scale, which was some 300 million years ago). The Mississippian culture corresponded with the Cahokia Mound Builders mentioned above.

Discussion in Comstock's *Geology*, 1844, regarding these tracks.

Thunderbird: Thunderbirds were significant icons for Native Americans living in the Midwest. These are part of a petroglyph display and preserve at Washington State Park in eastern Missouri, a site believed associated with Cahokia Mounds. Stone inscriptions are found over many parts of the U.S.; however, more are found in the western states as the drier climate protects them better.

Thunderbird-II: Same location as shown previously.

Group of potato-looking maces: Same location as shown above.

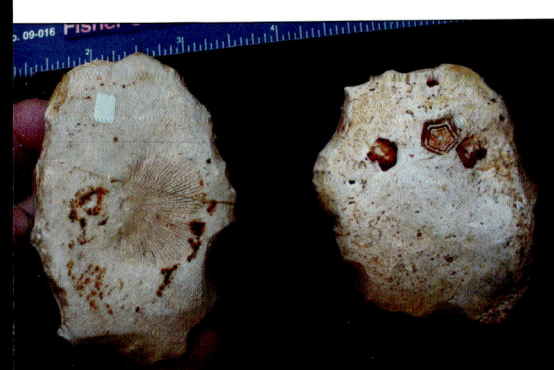

Worked (or knapped) flints from Bishops Spring, Dade County, Missouri. *Left*: brachiopod; *Right*: impression of crinoid plate. Mississippian (Late Paleozoic) chert and flint of the Ozarks of Missouri, Illinois, and Arkansas supplied Native Americans with the raw material for most stone tools in the southern mid-continent of North America. These worked flints were found along a spring branch of a medium sized Ozark spring associated with numerous flint chips—obviously Indians had camped by the spring and engaged in considerable flint knapping at the site. A Native American who, in some way, was curious about the fossil impressions he saw occupying the center of the knapped pieces made these particular knapped flints. Perhaps the knapper was a latent fossil collector who had to leave his specimens behind when the tribal group moved to another location and/or perhaps the knapper felt some religious significance associated with the fossils. We will never know but we can surmise that curiosity of the maker of these knapped flints was aroused by the fossils, just as you are while looking at this book.

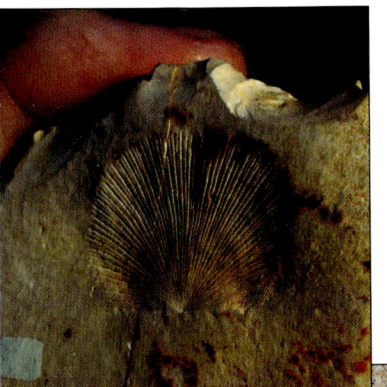

Close-up of brachiopod in flint of Mississippian age, which has been knapped at the edges with the fossil emplaced in the center. From the same locality as the above specimen in Dade County, Missouri.

Bison bison: This is the species of bison that lives today. These skulls are found on gravel bars and also in gravel digging operations on the Missouri and Mississippi rivers. In this case, the skull may or may not be a fossil, depending upon its age. *Bison bison* lived during the Pleistocene Epoch; however, unlike extinct species such as *B. latifrons*, *B. bison* survived the Pleistocene extinction event and produced the giant herds that supported the Plains Indians. It is difficult to determine if a skull like this is from the Pleistocene, and hence is a fossil (the boundary between the Pleistocene and the Holocene generally being placed at 10,000 years before the present), or is younger than 10,000 years and is from the Holocene, where it would not be old enough to be a fossil. Many of the bones and skulls found in sediments of the large inland rivers of the U.S. are not mineralized so they don't look like fossil bone.

Close-up of crinoid calyx plate in knapped flint from Dade County, Missouri.

A white (albino) buffalo (*Bison bison*) of the plains: Painting by Louis Shimschee, a Native American artist who painted animal and Native American scenes in the 1930s, '40s, and early '50s. Shimschee painted a variety of western scenes for tourists driving Highway 66 through Oklahoma where his works often were sold at roadside stands. The bison is one of the large animals of the Pleistocene Epoch that survived the extinction event; it then multiplied and produced the large herds that supported the Plains Indians.

Rain god: effigy from southeastern Missouri boot-heel obtained from a roadside stand in 1935 by Leonard and Virginia Stinchcomb. Such figures are associated with the extensive Mississippian Culture of Native Americans, a culture that reached its peak around 800 years ago and had its metropolitan center at Cahokia, western Illinois.

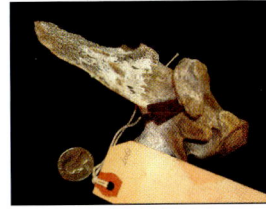

Bison vertebrae with embedded iron arrow or spear point: Bison bones with flint arrowheads embedded in them are sometimes found. These bones can be 3,000 to 300 years old, depending upon the type of projectile found in the bone. Iron arrowheads made by Native Americans replaced flint ones in the nineteenth century, iron being available from settlers, as well as railroad iron, etc. Such arrowheads ceased to be made by the twentieth century. Guns were easier to use and the buffalo had been decimated.

Broken arm of above "rain god:" Ceramic pottery of Native Americans often survived over time as the firing process stabilized the artifact's clay so it was able to withstand the degradation of time. The section of fired clay exposed in the broken arm of this effigy indicates a firing temperature of around 1,200 degrees F was reached, not high enough to remove all organic material from the interior of the effigy's arm, which has retained its original black organic material as can be seen in this cross section.

Discussion of bison herds leading to their possible extinction from Comstock's *Geology*, 1844: The author predicts that before the end of the nineteenth century the bison might be exterminated and become known only as a "fossil species." Note that the author of this paragraph defines a fossil species differently from that used by the author of this book. In Comstock's *Geology*, a fossil is considered to be any extinct life form—hence by his

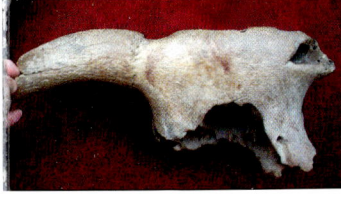

Ox skull: A variety of skulls (both fossil and modern) are found on gravel bars and in gravel digging operations on the Missouri and Mississippi rivers—often these skulls lack mineralization. In the case of this one, it probably isn't a fossil. It is (presumably) not old enough. This skull may have been from an ox used to pull wagons in the late eighteenth or nineteenth century in the St. Louis area where it was found.

definition, the Dodo, which went extinct in the late 1600s, would be a fossil species. The definition preferred by the author of this book is that any organism 10,000 years or older is a fossil—younger than that it is not, irregardless whether it is rock-like or not. Some modern authors use 8,000 years rather than 10,000 as the cut off point as to what is and is not a fossil.

Aepyornis madagascariensis: Elephant Bird Egg. A reconstructed egg of a giant bird called the elephant bird, which lived in Madagascar until the early 1600s, when it went extinct. These giant birds produced the largest egg of any known animal, exceeding in size even the eggs of dinosaurs, some of which were the size and shape of small watermelons. The egg of the elephant bird was even larger. An ostrich egg is being examined by Snuggles. Today the ostrich egg is the largest egg of any living animal.

Lithic arts-1. A craft that has become popular with some persons is the creation of modern flint tools from the same rock as was used by Native Americans in the past. These are such modern crafted stone "tools."

Another group of modern crafted stone tools: Persons who make these do so as an art form or craft. They sign their work and most have absolutely **no intent** of passing such work off as real artifacts. Some persons have severely criticized this activity as being totally wrong. It is a craft that persons should have as much right to engage in today as was done thousands of years ago. What is wrong, and fraudulent, is to try to pass off such items as being artifacts.

Stromatolite: This is a slice through a modern (Holocene) stromatolite. Stromatolites represent the oldest direct evidence of life on the earth, the oldest stromatolites occurring in rock strata 3.5 billion years old. Stromatolites are sedimentary structures produced by the life activities (primarily photosynthesis) of cyanobacteria or the blue-green algae as well as some other monerans such as photosynthetic bacteria. The specimen shown here is part of a stromatolite which formed in shallow water a few thousands of years ago in northwestern Australia near an area where they are still forming (and living), a region known as Shark Bay. (Value range E).

Group of modern stone tools and the flint blanks from which they are made.

A Big Question!

No topic concerning the history of life can be more significant than that regarding the appearance of humans and intelligence. This matter in science is relegated, as are all issues regarding "ascent" (or descent) in the development of life **to evolution**—and today especially to that method of evolution espoused by both Charles Darwin and Alfred R. Wallace in the mid-nineteenth century—**natural selection**.

The appearance of intelligence in humans carries with it a complex, interesting, and controversial legacy of questions and explanations involving both religion and science. Vast amounts of ink have contributed to the extensive literature written on this matter and such contemporary issues like modern genetics and evolution contrasts with compelling arguments regarding intelligent design and the nature of consciousness—the latter for which science currently seems to lack any explanation.

With regard to geologic time and the fossil record, modern man clearly is a late comer to the arena of life. The appearance of tool making (an activity indicative of a level of intelligence beyond that of any other life forms to have ever lived) clearly appears deep in the Pleistocene and less clearly (the tools are very crude and are barely tools at all) in the late Pliocene.

Some "Artifacts" of the Geology vs. Theology Controversy of the Nineteenth and Early Twentieth Century

The matter of appearance of intelligence in humanity, as stated above, is well beyond a work such as this; however, some particularly interesting "artifacts" relevant to this conflict exist both in early geology texts as well as in early sound recordings to which the author also confesses to having an interest. One of the more accessible and interesting of such recordings are those associated with the legacy of William Jennings Bryan, three times presidential nominee at the close of the nineteenth century and the beginning of the twentieth. This time period coincides with contemporary fear concerning the social consequences of the theory of evolution, a matter of particular concern at that time. Issues regarding Creation vs. Evolution, especially in the U.S., became strong enough so that many states, especially those of the old Confederacy, passed legislation which forebade the teaching of evolutionary theory in the public schools. Charles Darwin, in his 1859 *Origin of Species*, had nailed down a mechanism (natural selection) that could scientifically explain evolution, a topic that had been smoldering in Western Europe since the late eighteenth century. Late nineteenth and early twentieth century writings on the evolution vs. creation issue are relatively common, and they are interesting—audio artifacts on this topic are considerably rarer and tend to focus on those made by W. J. Bryan who incidentally was the prosecuting attorney in the famed Scopes "Monkey Trial" of 1925. With regard to this issue, it is incumbent to mention the numerous writings on this subject by the late Stephen J. Gould, especially with respect to Bryan. A significant portion of the persona of Bryan was framed in his strong belief in a creator and the teachings of Jesus as outlined in the New Testament. A liberal reformist, Bryan's political infrastructure was supported by the beautiful elegance of his oratory as preserved in some of these old recordings. It might be pointed out that Charles Darwin also foresaw problems that his theory might (or would) produce in social matters and with society in general—but he chose not to get involved. Some have suggested that his many ailments later in life may have been a defense mechanism to avoid these controversial issues. At any rate, the peak of the evolution-creation controversy was at the end of the nineteenth century and early in the twentieth and Bryan was a major player in this controversy on the side of creationism.

The items shown here (obviously not fossils as are none of the objects of this chapter—they are not old enough) are not only collectable, but it appears that little awareness of them exists outside of the "discophile collecting community," a collecting community which otherwise remains entirely separate from that community with a bent for fossils.

Just as sure as tomorrow there will be a geologic future—just as there has been a geologic past. Will the geologic future of a million years (not much time at all to the earth or even to its rock strata and fossils) see the existence of man? Will the earth still be a beautiful place or will it be a nether world of overpopulation or totalitarian despotism or will man have learned how to really live with such a unique planet? Will the geologic future find humans (or their chip-based equivalent) inhabiting much of the Solar System? Will the geologic future of 10 million years find humankind inhabiting distant parts of the Milky Way Galaxy—perhaps even as a hybrid between our carbon based selves and a form of consciousness inscribed on a silicon or other crystalline wafer? Are some of the enigmatic and elusive phenomena broadly dubbed as UFO's really us? Are they humans for whom the secrets of time travel have been discovered in this future? The future itself is elusive—the geologic future especially so with regard to the "geologically new" phenomena of intelligence—but there will be such! The phenomena of intelligence capable of contemplating itself has been a void until now—may it have a benign and awesome future!

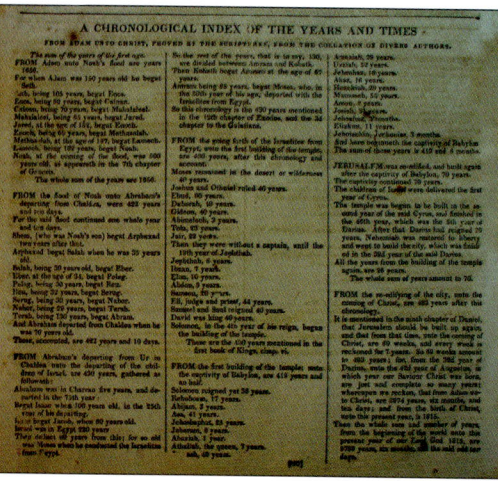

"Young earth Creationist's evidence:" Discussion of the time interval since the Creation of the earth and the present being 6,000 years. Prior to the 1830s, the biblical six days of creation was taken literally by many scientists. This was before a clear concept existed as to how earth processes acted upon the earth over vast time periods and the development of what is referred to as geological uniformitarianism. Early nineteenth century geologists (which included Charles Lyell) emphasized the link between strata, fossils, and megatime, ushering in modern geology. With this sort of geology, a conflict quickly developed in the early nineteenth century between this new science and some Christian theology. In the seventeenth century an Anglican clergyman, Archbishop James Usher determined on the basis of Scripture that the earth was about 6,000 years old. Usher's determinations being essentially what is shown here, where this particular account brackets the time from Adam and Eve until the birth of Christ as 3974 years (bottom right). The account then more correctly states the time **since** the birth of Christ to the present as being 1815 years. (1815 being the year this particular account was first published). Incidentally this account came from the family bible of the author's great-great-great grandfather, Noah Stinchcomb. It was published in Boston in 1826.

William J. Bryan's Cross of Gold speech on an 1897 cylinder record recording: William J. Bryan was a strong proponent of a literal interpretation of the Bible and a staunch Creationist. He ran for president in 1896 against William McKinley on the topic of the speech recorded on the cylinder, an issue involving the free coinage of silver, and against the backing of currency with gold. A religious motif shows up here in the speech's closing with the words, "You shall not press down upon the brow of labor this crown of thorns, you shall not crucify mankind upon a cross of gold."

Geology vs. Theology discussion, Comstock's *Geology*, 1844.

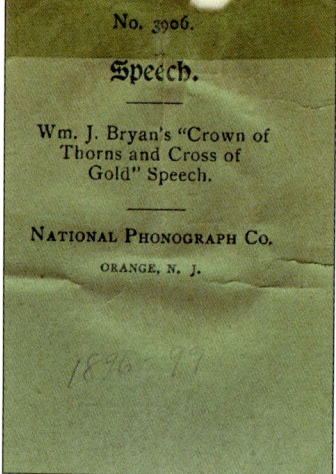

Record slip from Bryan's famous Cross of Gold Speech.

A proposal for reconciliation of Geology and religion in the mid-nineteenth century. The Dr. Buckland mentioned here was William Buckland, an Anglican clergyman who became one of the pioneers of modern geology. Buckland was a close friend of Charles Lyell of London.

Imperialism by W. J. Bryan: A 1904 cylinder recording opposing the annexation of the Philippines, spoils taken as a consequence of the Spanish American War. Bryan was a progressive liberal for his time, always championing the little guy and the "underdog" and, in this case, against the extension of an imperialistic empire in the Philippines.

Immortality by William J. Bryan: A recorded excerpt of a speech given widely by Bryan, dwelling with the uniqueness of man as compared to other forms of life, a position taken also by Alfred R. Wallace, the co-discoverer with Charles Darwin of natural selection as the mechanism of evolution.

The John T. Scopes Trial, Vernon Dalhart: An early "country and western" song dealing with the famous Scopes Trial of 1925. Bryan was prosecuting attorney for the state of Tennessee in this so-called "Monkey Trial" at Dayton, Tennessee. Bryan died shortly after the trial of a heart attack—the event being portrayed heroically in the 1952 movie *Inherit the Wind*.

Rock of Ages.

Natural phenomena, such as fossils or caves, can really be awesome! We don't want a future where (as is the case with this cave) it has been supplanted totally by manmade phenomena or placed off-limits. Natural phenomena are every person's inheritance—a wise balance between utilization and conservation has to be struck (*Troop 64, BSA, Ferguson Missouri, 1959*).

Bibliography

Gould, Stephen J., 1983. "A Visit to Dayton and Moon, Mann, and Otto" in *Hens Teeth and Horses Toes—Further Reflections in Natural History*. W. W. Norton and Co., New York-London. ISBN-0-393-30200-8.

_____. 1991, "William Jennings Bryan's Last Campaign" in *Bully for Brontosaurus. Reflections in Natural History*. W. W. Norton and Co., ISBN 0-393-02961-1

Glossary

Alluvial sediments: Sediments composing the floodplains of streams. Alluvial sediments are generally Pleistocene or Holocene in age and may contain the bones and teeth of Pleistocene or Holocene animals.

Asteroid: An irregular rock-like body smaller than a planet that occurs primarily between Mars and Jupiter in the Asteroid Belt. Asteroids, when they collide with each other, are believed to be the source of material responsible for impact craters like the Steinheim Structure of Germany, as well as numerous craters seen on the Moon and the terrestrial planets.

Asteroidal impact: An impact on a planetary surface by an asteroid that produces a crater, which can be tens of kilometers in diameter. Considerable disturbance to underlying rocks will accompany such an impact, which can include the formation of a central uplift as a consequence of rebound.

Astroblem: An impact crater that has been eroded from weathering on the earth. Most large impact sites on the earth of any geologic age have been influenced extensively by atmospheric processes and become astroblems.

Bedrock: Underneath the soil or regolith is rock! Such hard rock that lies below soft, surficial material is bedrock.

Breccia: Rock made up of angular fragments known as clasts. Breccias can be produced by a variety of means, which include impact, faulting, and other high-energy phenomena.

Calcareous: A rock or geologic material containing calcite or calcium carbonate.

Chalcedony: A finely crystalline form of quartz which exhibits a waxy appearance. Chalcedony is found to compose the silicified corals of Miocene and Pliocene age from Florida.

Champlain sea: A shallow sea that covered parts of Vermont and southern Quebec during the Pleistocene Epoch. The weight of continental glaciers depressed continental crust below sea level. When the glaciers melted, this depressed land was invaded by the sea, becoming the Champlain Sea, which is named after Lake Champlain of Vermont.

Clastic sediments: Sediment such as sand or gravel made up of fragments of previously existing rock material. This also includes finer grained sediment like clay and silt.

Clay lense: A layer of clay which is generally interbedded within some coarser clastic sediment like sand or gravel and often was the site of a former lake or lagoon. Such a clay layer is called a lense because the mass as a whole is often lens-shaped. Clay lenses can sometimes yield well-preserved plants and vertebrate fossils.

Compression fossil: A fossil formed when an organism (or part of an organism) is embedded and flattened in sediment and also where some of the original material is preserved in association with this impression. Fossil leaves are often preserved in this way.

Concretion: Round rocks that occur in shale, marl, and clay beds. Concretions in Cenozoic marine rocks can sometimes contain fossil crabs and lobsters, as the carapaces of these animals for some reason tend to promote forming a calcareous mass around them.

Consolidated sediment: Sediment which has become solid sedimentary rock. Cenozoic rock is often poorly consolidated and soft unless involved in tectonic activity such as that of the California coastal ranges where it then can be quite solid. Some of these sediments are also of deep-sea origin, where they today compose the sea cliffs and coastal mountains of that state.

Continental glacier: A large accumulation of ice that covers part of a continent (in contrast to a mountain or alpine glacier, which is much smaller in area). A glacier can move slowly and, in the process, sculpt underlying rock. This happened on a large scale during the Pleistocene ice age producing distinctive landscapes.

Copal and copalite: Copal is a terpinoid resin derived from tropical trees, usually the araucaracea. Copal is relatively soft when very fresh and may even occur as a very viscous liquid, which can flow and cover things, including insects. Copalite is hardened copal (or polymerized copal) that may occur associated in soil layers beneath copal producing trees. Copalite is geologically young, that is Holocene, Pleistocene or possibly Pliocene in age. Both copal and copalite were previously used extensively for compounding varnish.

Deep sea sediments: Sediments (or sedimentary rock) deposited on the sea floor of the open ocean in contrast to that deposited on continental shelves or on continental crust.

Denali Fault: A major fault (or fault system) in Alaska, the downthrown side of which can include sediments of Neogene age.

Down dropped side of a fault: A fault is a crack in the earth, often involving vertical motion. The down dropped side of a fault can either accumulate or preserve younger rocks. The opposite side of a fault, the up thrown side, usually has older rocks on it, the once overlying younger rock on that side having been eroded away.

Drift, Glacial: A general term for glacial sediments. The term "drift" is a relic from belief that this unsorted material was from the flood of the Bible's Old Testament.

Earthquake, great one of 1811-'12: A large earthquake in the early nineteenth century occurred in southeastern Missouri (Bootheel region) that affected drainage and topography. Charles Lyell mentions this event in reference to Pleistocene loess outcrops near Natchez, Mississippi.

Flint and chert knapping: The act of hitting and flaking hard, brittle rock to form some sort of stone tool. During the past two decades knapping has turned into a craft pursued by a number of persons. It is referred to as "lithic art."

Foraminifera: Protists which possess a (usually calcareous) covering called a test. The tests of foraminifera (forams) can be common fossils in marine Cenozoic rocks, but are usually to be seen only with use of a microscope. Some forams however are megafossils, with some of the largest being found in Cenozoic marine limestones.

Fossil: Any structural evidence of life (impressions, actual material or even single cells) from the geologic past—the geologic past being defined as time prior to 10,000 years ago.

Fossil resins: This includes a number of organic substances produced by the exudations of (usually tropical) trees and preserved in sedimentary rocks or as accumulations in soil from growth of ancient resin producing trees. Included are amber, burmite, Kauri gum, and copalite.

Ice age: That recent portion of earth's history when continental glaciers (as well as enlarged alpine glaciers) covered large parts of the continents—especially those of the Northern Hemisphere.

Igneous intrusion: Cooled, but once molten rock which at some time during the geologic past was injected into parts of the earth's crust. Geologic processes of mass wasting (primarily weathering and erosion, over spans of geologic time) can erode whatever covered the intrusion so that today the (intrusive) igneous rock formed from the intrusion will be at the earth's surface.

Indurated: Term used to indicate that the grains of a sediment or sedimentary rock have been well cemented together to make it hard. Neogene sedimentary rocks are usually soft and not well cemented together and thus are generally non-indurated or poorly indurated.

Lithified sediment: Sediment like limestone or sandstone that has been cemented or compressed together to form solid rock. Older sedimentary rocks are usually hard and lithified—those of the Cenozoic generally less so. Very soft Cenozoic sediments are often **non**-lithified and non-indurated—that is they are soft and often crumbly.

Loess: Silt sized, fine sediment usually derived from rock flour produced by the abrasive action of glaciers. Loess is usually of Pleistocene age and locally it can contain fossils, including the remains of ice age mammals like mammoths and mastodons.

Marl: Soft, calcareous rock or sediment often containing the fossil shells of mollusks. Marl is especially characteristic of late Cenozoic marine sediment.

Meteoroid: A space rock that, if it were to enter the earth's atmosphere and not be destroyed, would then become a meteorite.

Monomict breccia: A breccia made up of fragments derived from the same source. Mentioned here in reference to the limestone breccias of the Miocene Steinheim Astroblem.

Moraine, terminal: Poorly sorted sediment deposited at the end of a glacier.

Moraine, ground: Unsorted gravel, clay, and larger rocks that were overridden by movement of the glacier that formed and originally deposited them.

Paleokarst: An ancient (usually pre-Pleistocene) sinkhole or solution structure filled with sediment or sedimentary rock younger than that from which the sinkhole itself was formed. It's a kind of "fossil sinkhole."

Paleosoil: Ancient soil which is now part of a geologic section of rock strata and often has become lithified or turned to rock.

Permafrost: Permanently frozen soil or ground found in areas of high latitude. Permafrost beds can contain well-preserved bones (sometimes with fleshy remains as well) of mammoths and other ice age animals.

Polymerization: The chemical process or reaction by which small molecules (usually of an organic compound) become bonded or linked together to produce a solid material. Plastics are the best-known example of polymers, however amber and other fossil resins are natural polymers in which the resin molecules become polymerized with the passage of geologic time.

Probosidian: A member of the elephant family, which includes the extinct (late Cenozoic) mammoth, mastodon, and stegomastodon.

Pseudomorph: Literally "false form:" A mineral crystal that resembles that crystal-form representative of another mineral. Specifically in reference to crystals that resemble those of pyrite found composing iron-oxide-replaced coprolites occurring in Cenozoic strata of Oregon, Washington State, and the island of Madagascar.

Regolith: Loose material on the surface of a planet. Regolith on the earth is the mineral component of soil; the other component is humus that is organic and found, of all the planets of the solar system, only on the earth.

Residual clay: Clay left behind from solution or chemical weathering of rock—which often can be limestone. In this work, the reference is to red clay produced in the Caribbean by weathering and incorporated within limestone breccias containing the fossil shells of large land snails.

Rocks, specifically to those of Neogene age: Neogene rock is often not really hard rock at all, but rather sediment. The formation of sedimentary rock from sediment is a product of geologic time, and in most cases Neogene rocks are not of sufficient geologic age for the lithification process (rock making process) to have occurred. There are exceptions to this, particularly in regions of the earth where these younger rocks have been subjected to tectonic activity. Usually these geologically young rocks will be folded or involved in some way with earth movements. This often makes them hard and indurated, like those of earlier periods of geologic time.

Rock flour: Finely pulverized rock that can be produced by a variety of phenomena. In this work, rock flour is mentioned with regard to its being produced with a high velocity impact associated with the Steinheim Astroblem. It is also mentioned associated with rock abrasion resulting from continental glaciers, where rock flour is the parent material of loess.

Search image: An image that is in a person's mind while in the process of looking for something specific, like fossils.

Soil: Loose, surficial material on the surface of the earth. Soil is composed of **two** components, **humus** and **regolith**. Humus is derived from organic material, generally plant debris derived from leaves and sticks. Regolith is the mineral component of soil. As far as is known, true soil **only** occurs on the earth. To use the term soil for loose material on another planet, like Mars or the Moon, viz. Lunar soil, is erroneous as there is no component of humus in the surface materials of these planets—their surface material rather should be referred to as regolith.

Speliothems: Cave formations made from dripstone—a secondary form of calcium carbonate (limestone) associated with caves. Among the best-known types of speliothems are stalactites, columns, and stalagmites.

Sub-fossil: A fossil-like object derived in some manner from an organism which is less than 10,000 years old and is from the Holocene Epoch of the Neogene.

Tectonic activity: Having to do with earth movements such as mountain building, uplift, and faulting. earthquakes are a consequence of tectonic activity. With regard to recent geologic time, some regions of the earth, such as western North America and Alaska give evidence of recent extensive tectonic activity or tectonism in the form of strata of young geologic age (Pliocene and Pleistocene) being deformed and the layers tilted.

Tectonism: Having to do with tectonic activity.

Test (Testacea): The "shell" or covering of a single celled, eukaryotic organism (protist).

Tuffaceous: Containing tuff or volcanic ash—material derived from volcanic eruptions in the past.